Springer Theses

Recognizing Outstanding Ph.D. Research

Aims and Scope

The series "Springer Theses" brings together a selection of the very best Ph.D. theses from around the world and across the physical sciences. Nominated and endorsed by two recognized specialists, each published volume has been selected for its scientific excellence and the high impact of its contents for the pertinent field of research. For greater accessibility to non-specialists, the published versions include an extended introduction, as well as a foreword by the student's supervisor explaining the special relevance of the work for the field. As a whole, the series will provide a valuable resource both for newcomers to the research fields described, and for other scientists seeking detailed background information on special questions. Finally, it provides an accredited documentation of the valuable contributions made by today's younger generation of scientists.

Theses may be nominated for publication in this series by heads of department at internationally leading universities or institutes and should fulfill all of the following criteria

- They must be written in good English.
- The topic should fall within the confines of Chemistry, Physics, Earth Sciences, Engineering and related interdisciplinary fields such as Materials, Nanoscience, Chemical Engineering, Complex Systems and Biophysics.
- The work reported in the thesis must represent a significant scientific advance.
- If the thesis includes previously published material, permission to reproduce this must be gained from the respective copyright holder (a maximum 30% of the thesis should be a verbatim reproduction from the author's previous publications).
- They must have been examined and passed during the 12 months prior to nomination.
- Each thesis should include a foreword by the supervisor outlining the significance of its content.
- The theses should have a clearly defined structure including an introduction accessible to new PhD students and scientists not expert in the relevant field.

Indexed by zbMATH.

Youhong Chen

Stability Assessment of Power Systems with Multiple Voltage Source Converters

Bifurcation-Theory-Based Methods

Doctoral Thesis accepted by
The University of Manchester, Manchester, UK

Springer

Author
Dr. Youhong Chen
Department of Electrical and Electronic Engineering
The University of Manchester
Manchester, UK

Supervisor
Dr. Robin Preece
Department of Electrical and Electronic Engineering
The University of Manchester
Manchester, UK

ISSN 2190-5053 ISSN 2190-5061 (electronic)
Springer Theses
ISBN 978-3-031-63094-1 ISBN 978-3-031-63095-8 (eBook)
https://doi.org/10.1007/978-3-031-63095-8

This Springer imprint is published by the registered company Springer Nature Switzerland AG
The registered company address is: Gewerbestrasse 11, 6330 Cham, Switzerland

To my family

Supervisor's Foreword

The landscape of electrical power systems is undergoing a rapid transformation characterised by a significant surge in the deployment of power electronic converters. This surge has brought about fundamental shifts in the operational dynamics of our power systems. Transitioning from the era of synchronous generators, we are now entering an age dominated by voltage source converters (VSCs). However, this paradigm shift necessitates a swift evolution in our analytical tools and techniques.

Presently, we find ourselves inadequately equipped to comprehensively explain the intricate interactions and potential interferences among these VSC-based devices. Unforeseen operational issues have already occurred, contributing to instances of disruptive blackouts. This book addresses this critical gap by offering the essential analysis methods that are crucial to approach these complex challenges, thereby laying the groundwork for future breakthroughs in this domain.

While the mathematical framework employed—specifically, bifurcation analysis—may initially appear intricate, it serves as a valuable toolkit. This toolkit unveils previously unrecognised instability mechanisms, providing the broader research community with the means to establish operational guidelines and limits. These guidelines are instrumental in fully harnessing the capabilities of VSCs, mitigating the risks associated with potential instabilities.

This book presents considerable scholarly contributions, making it a valuable resource for those engaged in this field of study. The content is not only accessible but also methodically organised, facilitating an effortless understanding of the complex concepts presented. I look forward to seeing the progress of the research field following Dr. Youhong Chen's research.

Manchester, UK
June 2024

Dr. Robin Preece

The original version of the book has been revised. A correction to this book can be found at https://doi.org/10.1007/978-3-031-63095-8_8

Parts of this thesis have been published in the following journal articles and conference proceedings:

International Journal Publications

[J1] **Y. Chen**, R. Preece, and M. Barnes, 'Improving system loadability with the consideration of multiple bifurcations', International Journal of Electrical Power & Energy Systems, vol. 132, p. 1045, Nov. 2021.

[J2] **Y. Chen**, R. Preece, and M. Barnes, "Identifying Feasibility Region Boundary in Power Systems with Multiple VSCs," IEEE Trans. Power Syst., vol. 38, no. 2, pp. 1229–1241, Mar. 2022.

[J3] **Y. Chen**, L. Benedetti, R. Preece, P. N. Papadopoulos, M. Barnes, and A. Egea-àlvarez 'Investigating small-disturbance stability in power systems with grid-following and grid-forming VSCs using hybrid modelling approaches', Electr. Power Syst. Res., vol. 211, p. 108448, Oct. 2022.

[J4] **Y. Chen**, R. Preece, and M. Barnes, 'A Framework for Analyzing System Loadability with Multiple VSCs using a Hybrid Model', IEEE Trans. Power Syst., vol. 39, no. 1, pp. 1079–1094, Jan. 2024.

International Conference Publications

[C1] **Y. Chen**, R. Preece, and M. Barnes, 'Nonlinear Analysis of SSR in Wind Power Plants', in 16th IET International Conference on AC and DC Power Transmission (ACDC 2020), 2020, pp. 1–6.

[C2] **Y. Chen**, R. Preece, and M. Barnes, 'Analyzing Controller and Device Interactions in Hybrid Transmission Networks with High Penetration of Renewable Energy—A Review', 2021 IEEE Madrid PowerTech, PowerTech 2021—Conference Proceedings, June, 2021.

Acknowledgements

I would like to express my appreciation to my supervisors Dr. Robin Preece and Prof. Mike Barnes for their invaluable guidance and advice over the past four years. Their continuous support and expertise have significantly enhanced the quality of this study to its best possible form. It has been a privilege and an honour working with them.

A special acknowledgement must also go to the Engineering and Physical Sciences Research Council (EPSRC) and UK Research and Innovation (UKRI) for sponsoring this project. I am particularly grateful to Prof. Peter Crossley for providing me with the valuable opportunity to join the Centre for Doctoral Training (CDT) for the Power Network project.

My heartfelt thanks also goes to my colleagues in the Power Systems Research Group at the University of Manchester and the University of Strathclyde. The opportunities to discuss and develop ideas with them have been instrumental in ensuring the success of this research. Special thanks goes to my colleagues, Surat Asvapoositkul, Zaichun Zhang and Yitian Dai, whose continuous support and rigorous discussions throughout my Ph.D. study have been invaluable.

I am also grateful to my family for their unwavering support and encouragement throughout my studies and life in the UK. Without their unconditional support, it would be impossible for me to get through this difficult period.

Declaration No portion of the work referred to in this thesis has been submitted in support of an application for another degree or qualification of this or any other university or institute of learning.

Contents

Notation and Abbreviations

Symbols

A, B, C, D	State-space matrices
B	Susceptance
C	Capacitance
C_{eq}	VSC DC side equivalent capacitance
$Cov(.)$	Covariance of the variables
D	Generator damping constant
E	Voltage
E'_d	Transient voltages in d-axis
E_{fd}	Field voltage
E'_q	Transient voltages in q-axis
E_t	Output voltage
e_d	VSC terminal voltage in the d-frame
e_q	VSC terminal voltage in the q-frame
$F(\cdot)$	Nonlinear differential-algebraic equations
$f(\cdot)$	Nonlinear vector field associated with the state variables
$g(\cdot)$	Nonlinear vector field associated with the algebraic variables
H	Generator inertia constant
I	Current/Identity Matrix
I_d	Currents in d-axis
I_q	Currents in q-axis
i_{dc}	DC current from the DC network
i_n	VSC output DC current
i_r	Receiving-end current
i_s	Sending-end current
M	The inertia constant of the machine
N	Transformer turns ratio/Number of
N_{opt}	The optimal number of corrector iterations
n	Power system order/Number of
K	Controller gain

K_a	AVR gain
K_e	Exciter gain
K_f	Stabilising feedback loop gain of AVR
K_i	Integral gain
K_p	Proportional gain
k	The direction of change of the bifurcation parameter
L	Inductance
l'	The real part of the left eigenvector
l''	The imaginary part of the left eigenvector
P	Active power
p	Control parameter
Q	Reactive power
R	Resistance
r	Limits in the system
r_s	Spearman correlation coefficient
T	Time constant
T_a	AVR time constant
T'_{d0}	Transient time constants in d-axis
T''_{d0}	Sub-transient time constants in d-axis
T_e	Exciter time constant
T_f	Stabilising feedback loop gain of AVR
T'_{q0}	Transient time constants in q-axis
T''_{q0}	Sub-transient time constants in q-axis
t	Time
V	Voltage
V_{dc}	VSC DC side voltage
V_d^{sg}	SG's terminal voltages in d-axis
V_q^{sg}	SG's terminal voltages in q-axis
v_r	Receiving-end voltage
v_s	Sending-end voltage
X	Reactance
$X_d{}'$	Transient reactance in d-axis
X''_d	Sub-transient reactance in d-axis
X_{ls}	Leakage reactance
$X_q{}'$	Transient reactance in q-axis
X''_q	Sub-transient reactance in q-axis
x	System state variable
Y	Admittance
y	Algebraic variable
z	Tangent vector
Z	Impedance
Δ	Small derivation
α	Reference angle of the global (DQ) frame
β	Reference angle of the local (dq) frame
γ	Tap ratio magnitude

δ	Generator rotor angle
ε	Error
θ	Voltage angle
λ	Complex eigenvalue
μ	Bifurcation parameter
o	Step size
σ	The real part of the eigenvalue
τ	Time constant
ψ	Left eigenvector
ω	The imaginary part of the eigenvalue
ω_r	Angular speed
ω_{ref}	System frequency
ϕ	Right eigenvector
θ_{PLL}	The angle measured by the PLL
θ_{shift}	Tap ratio phase shift
$\rho_{X,Y}$	The Pearson correlation between X and Y
$\rho_{X,Y,Z}$	The partial correlation coefficient between X and Y
$\psi_{1d,2d}$	Flux linkages in damping winding circuits

Subscripts

0	Initial values/equilibrium point
abc	Abc frame
ac	AC system quantity
b_n	Pair-wise co-primes
$best$	Best value
bou	Boundary
c	Critical value
D	D-axis (system reference frame)
d	d-axis (machine reference frame)
dc	DC system quantity
eq	Equivalent
ex	Excitation system
G	Generation (load flow)
i	Index value, where $i = 1, 2, 3\ldots$/Integral
j	Index value, where $i = 1, 2, 3\ldots$
k	Index value, where $k = 1, 2, 3\ldots$
L	Load (Load flow)
$line$	Lines (or branches)
max	Maximum
Min	Minimum
N	Number

p	Proportional
P	Power
PLL	Phase-locked loop
Q	Q-axis (system reference frame)/Reactive power
q	q-axis (machine reference frame)
s	Shunt connected (Load flow)
sg	Synchronous generator
sys	System
V_{ac}	AC voltage
V_{dc}	DC voltage
vsc	Voltage source converter

Superscripts

.	Derivative
$'$	Transient/Real part
$''$	Sub-transient/Imaginary part
-	Normalised
*	Reference set-point
^	Estimated
~	Augmented
bus	Bus
dyn	Dynamic area
G	Synchronous Generator
j	Index value, where $j = 1, 2, 3\ldots$
$load$	Load
sg	Synchronous generator
sta	Static area
T	Matrix transpose
vsc	Voltage source converter

Abbreviations

ABB	Asea Brown Boveri
AC	Alternating current
AVM	Average-value model
AVR	Automatic voltage regulator
BECCS	Bioenergy with Carbon Capture Usage and Storage
BPNN	Back-propagation neural network
CDT	Centre for Doctoral Training

CIGRE	International Council on Large Electric Systems
CPU	Central processing unit
CSC	Current source converter
DAE	Differential-algebraic equation
DC	Direct current
DDSSO	Device-dependent sub-synchronous oscillation
EMT	Electromagnetic transient
EPSRC	Engineering and Physical Sciences Research Council
FACTS	Flexible AC Transmission System
FFR	Fast frequency response
HB	Hopf bifurcation
HVDC	High-voltage direct current
IEEE	Institute of Electrical and Electronic Engineers
LCC	Line commutated converter
LIB	Limit-induced bifurcation
LIDB	Limit-induced dynamic bifurcation
LISB	Limit-induced static bifurcation
MADM	Multi-attribute decision-making
MIMO	Multiple-input-multiple-output
MMC	Modular multilevel converter
MTDC	Multi-terminal direct current
NGET	National Grid Electricity Transmission plc
ODE	Ordinary differential equation
PC	Personal computer
PCC	Point of common coupling
PCM	Probabilistic collocation method
PE	Power electronic
PI	Proportional-integral (Control)
PLL	Phase-locked loop
POD	Power oscillation damping
PSO	Particle swarm optimisation
PSS	Power system stabiliser
PV	(Active) Power–voltage
PWM	Pulse-width modulation
QP	Quadratic programming
RES	Renewable energy resource
RGA	Relative gain array
RMS	Root mean square
RMSE	Root-mean-square error
SCC	Self-commutated converter
SCR	Short-circuit ratio
SG	Synchronous generator
SIB	Singular induced bifurcation
SIMC	Skogestad internal model control
SISO	Single-input-single-output

SM	Sub-module
SNB	Saddle-node bifurcation
SQP	Sequential quadratic programming
SRF	Synchronous reference frame
SSR	Sub-synchronous resonance
SVC	Static VAR compensator
TOPSIS	Technique for Order Performance by Similarity to Ideal Solution
UK	United Kingdom
UKRI	UK Research and Innovation
USA	United States of America
VSC	Voltage source converter
WSCC	Western Systems Coordinating Council

Chapter 1
Introduction

This research focuses on using bifurcation theory to identify and analyse system stability limits in power systems with power electronics devices connected. Renewable energy sources (RESs) are increasingly being connected to power systems to address the increasingly severe global warming problem and other related environmental issues [1–3]. Consequently, a large number of power electronics devices will be required to effectively integrate RESs into future power systems. Therefore, power electronic-interfaced components are highly likely to be widely used in modern power systems for various applications, including high-voltage DC transmission (HVDC) and battery energy storage systems [4–7].

Due to the increasing load demand and the need to improve network utilisation, power systems are being operated closer to their stability limits. The use of power electronic-based devices to integrate RESs also requires the use of new controllers which differ from those traditionally used in power control systems, potentially threatening the power system stability. Due to the wide-scale application of converters, it is predicted that multiple power electronic-based devices will be closely connected to each other in the future. As the controllers of each converter will operate at similar speeds, the potential exists for interactions between the control loops of different devices [8]. This undesirable behaviour among power electronics may be detrimental to system stability, particularly in grids characterised by high impedance and low inertia (i.e. weak grids) [9]. This makes it important to fully explore system stability limits, and to take the impact of the converter and the interaction among power electronic-interfaced components into account.

Additionally, the controllers of power electronics operate over a wide range of bandwidths, ranging from a few hertz to several kilohertz. This means that power electronics-based devices can impact a wide range of dynamic phenomena, blurring the boundary between different types of stability [8]. This has led to a growing necessity to assess system operating limits while incorporating simultaneous consideration of multiple types of stability. Conventional analysis techniques typically focus on revealing the instability mechanisms of specific power system issues and can thus

Y. Chen, *Stability Assessment of Power Systems with Multiple Voltage Source Converters*, Springer Theses, https://doi.org/10.1007/978-3-031-63095-8_1

analyse system operation limits only with respect to a single type of stability at a time. The bifurcation approach, which enables the joint consideration of various power system instability is used in this thesis to thoroughly assess system stability limits.

1.1 Power System Stability

Power system stability refers to the ability of a power system to maintain stable operation based on given initial operating conditions and to regain an acceptable stable operating point following a physical disturbance [8]. Power systems are continuously subjected to small disturbances such as connection and disconnection of loads, and each system should be able to respond to these changes properly, without failure. As a power system grows in size, the system is more likely to suffer from such small disturbances.

There exist various forms of instabilities in the power system resulting in high dimensionality and complexity of stability problems. Proper classification of stability into appropriate categories can greatly contribute to the analysis of complicated stability issues [10]. Power system stability can be classified based on the physical nature of the system, the size of disturbances, and the timescale of the phenomena. Based on the work in [11], power system issues are thus broadly classified into three categories: rotor-angle stability, voltage stability, and frequency stability. This classification was defined by a joint Task Force formed by the International Council on Large Electric Systems (CIGRE) Study Committee and the IEEE Power System Dynamic Performance Committee. Since the definition published in [11] was created, Fig. 1.1 illustrates the overall picture of the power system stability problem, as published in [11], showing its categories and subcategories.

However, there has been a significant change in the characteristics of power system dynamics. Due to the increasing number of power electronics-based devices in modern power systems, the dynamic behaviour of electrical power systems has become predominantly characterised by the associated fast-response, converter-based devices. Accordingly, the power system definition was updated in 2020 by an IEEE Task Force [12]. Two new stability classes, converter-driven stability and resonance stability, were introduced into the updated definition to incorporate the effects of power electronics devices. Figure 1.2 shows the current overall classification of power system stability.

1.1.1 Rotor Angle Stability

Rotor angle stability describes the ability of the interconnected synchronous generators in the power system to maintain synchronous operation under normal operating

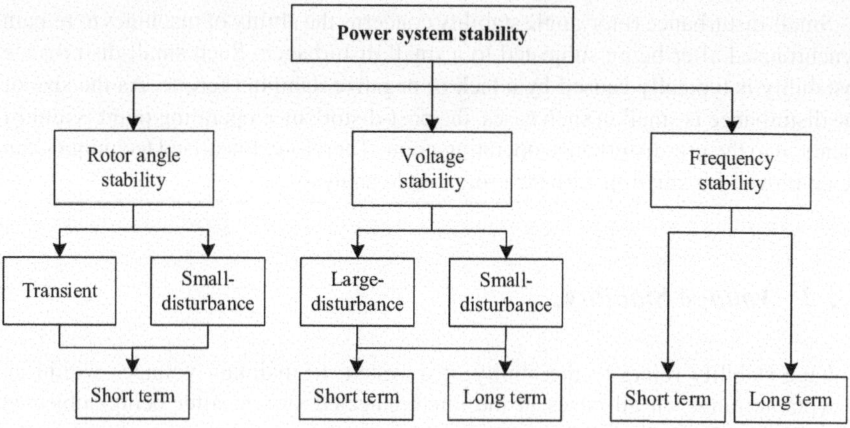

Fig. 1.1 Definition of stability of power systems published in 2004. © 2004 IEEE reprinted, with permission, from [11]

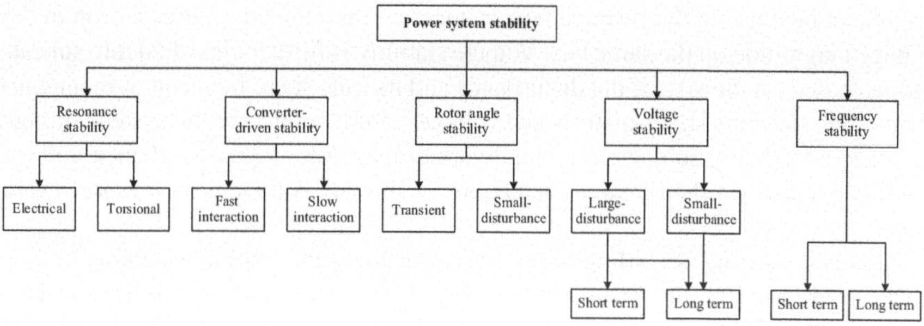

Fig. 1.2 Definition of stability of power systems published in 2021. © 2022 IEEE reprinted, with permission, from [12]

conditions and to regain synchronism after being subjected to small or large disturbances. In a synchronous machine, mechanical torque is associated with the power from the prime mover, while electromagnetic torque is associated with the electrical power from the generator terminals. A synchronous machine remains synchronised if these two torques are equal and opposite. Rotor angle stability is further divided into two sub-categories according to the size of the possible disturbance: large-disturbance rotor angle (transient) stability and small-disturbance rotor-angle stability.

Large-disturbance rotor angle (transient) stability concerns the ability of a machine to remain synchronised after being subjected to a severe disturbance. Insufficient synchronising torque leads to aperiodic or non-oscillatory transient instability. This type of instability is commonly related to large deviations in generator rotor angles. This is typically analysed via time-domain simulation, due to the nonlinear power–angle relationship that exists.

Small-disturbance rotor-angle stability concerns the ability of machines to remain synchronised after being subjected to a small disturbance. Such small-disturbance instability is typically caused by a lack of negative damping torque. As the size of the disturbance is small in such cases, the post-disturbance operating point is almost identical to the pre-disturbance operating point. Therefore, linearised techniques can be employed for small-disturbance rotor angle analysis.

1.1.2 Voltage Stability

Voltage stability refers to the ability of a system to maintain voltages within an acceptable range on all buses in the interconnected system after being subjected to a disturbance. Voltage stability is linked to reactive power balance between the load demand and supply. The relationship between reactive power and voltage is governed by Q–V characteristics. Under normal operating conditions, an increase in reactive power leads to an increase in bus voltage. However, when voltage instability occurs, an increase in the reactive power injection onto the bus causes a drop in the voltage magnitude on the same bus. Voltage stability is further classified into subcategories based on the size of the disturbance and its time span. In a similar manner to rotor-angle stability, small-disturbance voltage stability concerns the system voltage response after being subjected to small variations, while large-disturbance voltage stability focuses on the dynamic behaviour of the bus voltage after a more severe fault.

Alongside sorting according to the size of disturbances, voltage stability can be categorised based on the timescale involved. Short-term voltage stability involves the dynamics of fast power components such as induction motors and inverter-based generators, whereas long-term voltage stability involves slower-acting power devices such as tap-changing transformers. Long-term voltage instability usually results from the loss of the system equilibrium point as the load dynamics attempt to restore power consumption beyond the maximum power transfer limit. Based on this, long-term voltage stability is investigated by assessing the system stability limit in terms of increases in load demand.

1.1.3 Frequency Stability

Frequency stability is defined as the ability of a power system to maintain its operating frequency within an acceptable limit after being subjected to severe disturbances that cause a significant imbalance between the electrical supply and load demand. Frequency instability is typically caused by the slow response of controllers or a lack of generation reserves. Severe power system faults, such as short circuits on the transmission line due to lightning or the loss of large generation units, can result in large disruptions in the system frequency, which may cause a full system blackout.

Frequency stability can be further divided based on the response speed of control devices, resulting into two sub-categories short-term frequency stability and long-term frequency stability. This classification is determined by the varying response times of devices, which can range from a few seconds for short-term frequency stability to several minutes for long-term frequency stability.

1.1.4 Resonance Stability

Resonance refers to oscillations resulting from periodic exchanges of energy. These oscillations increase where there is excessive energy that cannot be dissipated in the power flow and may occur in voltage, torque, or current, often at increasing magnitudes [13]. Resonance instability occurs when the growing magnitude of such oscillations exceeds a specified threshold. The term sub-synchronous resonance (SSR) was defined in [8] to describe this phenomenon, although this is encompassed by the broader idea of resonance stability. SSR can manifest in two forms: torsional resonance and electrical resonance.

One type of torsional resonance is typically associated with a resonance between the network series compensation and the mechanical torsion of the turbine-generation shaft, which creates significant energy exchange between the network and shaft at sub-synchronous frequencies. Another form of torsional resonance is a device-dependent sub-synchronous oscillation (DDSSO), which results from the interaction of fast-acting control devices and torsional modes from nearby turbine generators.

Electrical resonance, also known as self-excitation SSR, refers to the resonance among series compensation devices and the electrical characteristics of induction generators. This results in large oscillations in current and voltage that damage the electrical power components.

1.1.5 Converter-Driven Stability

The dynamics of converter-based devices differ from those of conventional synchronous machines. High-level controllers of power electronics devices can typically operate in the range of 1–10 Hz, while the switching frequency of the converter operates at the kilohertz level. Due to the wide operating timescale of the converters, power electronics devices have the potential to interact with various phenomena, ranging from electromagnetic transients to voltage stability. Based on the frequencies of the observed power system issues, converter-driven stability can be further categorised as slow-interaction converter-driven stability (typically below 10 Hz) and fast-interaction converter-driven stability (typically from tens of hertz to kilohertz).

Fast-interaction converter-driven stability involves fast interaction between the controls of power electronics-based devices and fast-response power components such as the transmission network. Slow-interaction converter-driven stability is, in turn, associated with the slow dynamic interaction between the control systems of power electronics and the slow-response components of power systems such as the electromechanical dynamics of synchronous machines. Slow-interaction converter-driven instability shares a similar mechanism with voltage stability, which is mainly affected by the maximum power transfer capability between the converter and the rest of the system, particularly under weak grid conditions.

The aim of this thesis is to analyse various power issues using a bifurcation theory-based approach. Therefore, this thesis considers several different types of stability, including small-disturbance rotor-angle stability, voltage stability, electrical resonance stability, and converter-driven stability. It should be noted that this thesis takes the term small-disturbance stability to refer to stability that can be identified and analysed based on the modal analysis of the system Jacobian matrix. Specifically, small-disturbance rotor-angle stability, electrical resonance stability, and converter-driven stability are described using the general classification of small-disturbance stability within this thesis.

1.2 Voltage Source Converters

Due to the wide application of power electronic-interfaced devices, power system dynamics have come to be predominantly characterised by the effects of these components. Voltage source converter (VSC), a typical type of power electronics interface device technology, has a wide range of energy applications, such as HVDC and flexible AC transmission systems (FACTS), which can be employed in the various aspects of power systems including, electric power generation, transmission and distribution systems [14]. A large number of VSC-interfaced devices are thus expected to be integrated into future power systems.

Take the application of VSCs to renewable power generation as an example. With the severity of global warming being increasingly recognised, governments in most countries are now co-operating to reduce their respective carbon footprints to mitigate further increases in global temperatures. The UK government, for example, is committed to reducing its greenhouse gas emissions to net zero by 2050 [15, 16]. To achieve this goal, a massive amount of RESs are expected in future power systems. According to the Leading Way scenario (the most optimistic scenario) provided by National Grid's Future Energy Scenarios report [17], wind farm power generation capacity should reach 50 GW by 2030, and increase further through the 2030s. Clean energy including wind, solar, nuclear and Bioenergy with Carbon Capture Usage and Storage (BECCS), is expected to provide 94% of the country's power generation capacity by 2050. Even under the Falling Short scenario (the worst-case scenario), the capacity of wind power is expected to grow to 31 GW by 2030. By 2050, clean energy will contribute between 70% and 84% of the power generation capacity. The energy generation is characterised by renewable energy, particularly offshore wind, but conventional fossil fuels still play a vital role in power systems under the Falling Short scenario. Many RESs, including wind power, are integrated into networks by means of power electronic devices, so all of these scenarios suggest that converter-based devices will be increasingly connected to power systems.

Additionally, the increasing integration of VSC-HVDC techniques also means the wide application of VSCs in future power systems. Large offshore wind farm connections with capacities of roughly 500 MW and above are typically located more than 50 km away from the main grid. AC cables are not suitable for such long-distance, large-capacity power transmission as a significant amount of current is used in reactive power transfer to support the required voltage magnitude [7]. This leads to higher power losses and higher operational costs. Consequently, HVDC is the most feasible connection solution in these cases. Two main technologies are currently used in HVDC links: line commutated converter-based HVDCs (LCC-HVDC) and VSC-based HVDCs (VSC-HVDC). Conventional LCC HVDCs can provide bulk power transfer over long distances and were most commonly used in early-stage projects [4, 7]. VSC-HVDC technology is now considered the main candidate for the connection of large-scale RESs, particularly for offshore wind farm connections. Compared to LCC-HVDC, VSC-HVDC is preferred because of its smaller footprint, its black start capability, and its potential to support AC grids, particularly for weak AC systems, as it has the capability of controlling active reactive power independently [6, 18, 19].

Prior to 2012, there were four HVDC links existed in the UK. These interconnectors enabled a maximum of 3 GW of power transfer between the UK and mainland Europe, and 1 GW of transfer to the Island of Ireland with 0.5 GW of this using the VSC-HVDC technique. In the present day, the number of HVDC link projects built in the UK has increased to ten with a total capacity of 12 GW (3.9 GW from embedded links and 8.1 GW from interconnectors) [20]. Moreover, it is forecasted that a total of fourteen HVDC links, including interconnectors between the UK and Europe and embedded links within the UK power grid, are expected to be completed across the UK power system by 2030 [21]. In general, it is anticipated that VSC-HVDC will continue to be widely used in future power systems. This indicates a widespread proliferation of VSC technology in future power networks.

FACTS technology is referred to as a power electronic-based system combined with other static equipment that can control active or reactive power at the point of common coupling [22]. This enhances the system's controllability and increases its power transfer capability. The first generation of FACTS devices was primarily constructed using thyristor-switched capacitors and reactors. With the advancement of power electronic switches, a variety of new VSC-based FACTS technologies have emerged. [14]. FACTS devices are used to compensate for the intermittent characteristics of renewable energy sources. As discussed above, renewable energy will continue to replace conventional coal-fired generation. It is expected that FACTS will play a significant role in future power systems. This suggests numerous VSC-interfaced devices will be integrated into the next-generation electricity grid.

1.2.1 Historical Development of HVDC

The HVDC transmission was even well-known before the discovery of alternating current. The first trial DC transmission system was built in 1882. It was designed to transmit 1.5 kW over 25 miles. Even in the early twentieth century, several DC links, for instance, the 125 kV, 20 MW Moutiers–Lyon link (230 km), remained operational [23].

In 1939, Dr. Uno Lamm invented a new design for the mercury-arc valve, which greatly contributed to the development of HVDC conversion. This innovation led to the first commercial mercury arc-based HVDC system in 1954—the Gotland 1 link in Sweden, which transmitted 20 MW at 100 kV over 96 km.

By the 1970s, a solid-state technology, i.e. thyristor, was found as a better alternative for HVDC conversion and had started to replace mercury-arc valves used in early installations. Thyristors or Silicon Controlled Rectifiers appeared in the 1950s. However, the early-stage thyristors had problems with low electrical characteristics. The 1960s have witnessed the rapid development of thyristors. The blocking voltage of a single thyristor reached 1.6 kV and a maximum current at 1 kA. Compared with mercury-arc valve systems, thyristor valve-based HVDC systems have unique advantages of reducing maintenance, removing incidences of arc backs and extending voltage range. The first fully thyristor-based HVDC system was the Eel River scheme in Canada (1972), transmitting 320 MW at 80 kV between New Brunswick and Hydro Quebec. Both thyristors or mercury arc rectifier-based systems are called 'Current Source Converters' (CSC) or 'Line-Commutated Converters' (LCC) [24]. LCC systems rely on the line voltage of the connected AC system for current commutation. Such a system requires a strong AC system to undertake the conversion and a large amount of reactive power for support. Besides, it generates a large number of low-order harmonics resulting from the commutation process.

In the 1990s, the insulated-gate bipolar transistor (IGBT) became available, marking another evolution in HVDC technology. The HVDC systems based on IGBTs are referred to as 'Voltage Source Converters' (VSC) or 'Self-Commutated Converters' (SCC). Unlike LCC systems, VSCs can self-commutate and do not rely on a strong AC system. Therefore, they can switch at a high frequency, which could eliminate the low-order AC harmonics [24]. In 1997, the first experimental VSC-HVDC was tested in Hellisjon, Sweden, which had a capacity of 3 MW at 10 kV.

1.2.2 Current VSC Technology

In 1999, ABB built its first commercial VSC HVDC project, in Gotland, Sweden, which was a submarine cable transmitting 20 MW at 80 kV. It was a two-level converter based on the use of pulse-width modulation (PWM) to generate AC voltage [25]. By controlling the turn-on and the turn-off of the semi-conductor devices, the two-level converter in this approach simulates sinusoidal three-phase AC voltage and feeds it into the AC network. In ABB's second generation of VSC-HVDC technology, a three-level converter topology was used to reduce switching losses. A new harmonic cancellation scheme, Optimum PWM, was also then proposed by ABB to be adopted in the third generation of VSC-HVDCs to further reduce switching loss from 3.7% per converter when a two-level converter topology is employed to 1.8% [25]. The first VSC-HVDC project connecting offshore wind farms, BorWin1, used this type of technology.

VSC-related HVDC converter technology has been updated and developed by several different companies, including Siemens and ABB. HVDCs are now experiencing a shift from two-level converter topology towards multiple-level converter topologies. A VSC-HVDC based on a modular multi-level converter (MMC) topology was first used by Siemens in the Trans Bay Cable project in 2010. After that, the MMC-based VSC HVDC technology attracted the attention of both academics and manufacturers. By determining the appropriate combination of switching states for sub-modules in the MMC, a wide range of AC voltages can be provided, thus enhancing the quality of converter output voltage waveforms and substantially reducing the harmonic components in the output voltage. Furthermore, the use of MMC-based VSC HVDC reduces converter loss to as little as 1%, making it comparable with LCC HVDC technology. However, compared with two-level converters, MMC-based VSC HVDCs are more complex, which require more complicated insertion control of each sub-module [26]. Besides, MMC also suffers from additional power issues, including the unbalanced sub-module capacitor voltage [27] and circulating current between the converter limbs [28].

1.2.3 Impact of VSCs on System Stability

As significant numbers of VSC-interfaced devices are integrated into power systems, the dynamic behaviour of systems has become progressively dominated by fast-response power electronics devices. As discussed in Sect. 1.1.5, the timescale of the controller related to VSCs may vary from several microseconds to a few milliseconds. Such a wide timescale highlights the potential for VSCs to cross-couple with both electromechanical and electromagnetic phenomena. VSC-interfaced devices can interact with the dynamics of generators, transmission networks, and other power electronics components. These undesirable interactions may potentially degrade the system's stability. This has led to a growing interest in investigating the dynamics of these integrated VSC-interfaced components with respect to their impact on system stability. Section 1.5.1 will offer a review of existing research into this and other related areas.

1.3 Bifurcation

A power system is a highly nonlinear dynamic system that contains many nonlinear devices, including generators, power electronic-interfaced devices and protective devices. In any dynamic system, a qualitative change in the behaviour caused by changes in one or more parameters is recognised as bifurcation [29, 30]. Within power systems, the occurrence of bifurcation due to variations in system parameters is commonly associated with a complicated system response, which may potentially lead to system instability. Hence, the identification of these bifurcations in power systems has attracted increasing attention from researchers in recent years. Bifurcation theory provides a strategy for capturing and analysing critical bifurcations by identifying unique patterns of each type of bifurcation.

The following sections thus introduce the basic concepts related to bifurcation theory, including an introduction to dynamics, a definition of a fixed point with its stability, and the types of bifurcation of interest in this thesis.

1.3.1 Dynamic System

Although dynamics is now seen as an interdisciplinary concept, it was initially developed as a branch of physics. The concepts of dynamics were first introduced in the 1600 s, when Newton invented differential equations to discover the laws of motion and universal gravitation. Differential equations were then employed to explain Kepler's law of planetary motion. In the 1800s, Poincare developed a powerful new geometric approach based on differential equations that provided a qualitative rather than a quantitative solution to nonlinear dynamic problems, the latter of which forms the prototype for modern bifurcation theory analysis.

Differential equations can be used to describe system behaviour over a continuous time period, and have been widely used in science and engineering. Differential equations play a vital role in the modelling of virtually all physical processes, chemical reactions, and even business cycles, with their scale of application ranging from celestial movements to the interaction between atoms. In power systems, more specifically, differential equations are employed to model the dynamic responses of power components and the dynamic characteristics of systems.

In general, a system is considered linear if (and only if) the principles of superposition and homogeneity properties are valid [31]. Otherwise, the system is deemed nonlinear. Power systems are typical nonlinear dynamic systems and are thus commonly represented by sets of nonlinear differential equations.

1.3.2 Fixed Point

Consider a dynamic system represented as a differential equation in the form $\dot{x} = f(x)$. A graphic analysis of the dynamic system is shown in Fig. 1.3. The horizontal axis in Fig. 1.3 is x, denoted as the position of a particle moving along the real axis, and the vertical axis is $\dot{x}$, which is the velocity vector of that particle at each value of x. As shown in Fig. 1.3, the arrow on the real axis indicates the sign of the corresponding velocity vector. The arrow points to the right when $\dot{x} > 0$ and to the left if $\dot{x} < 0$. Based on this, assume there is a steady fluid flowing along the real axis following the direction of the arrows. The figure shows that the fluid flows towards the right when $\dot{x} > 0$ and towards the left if $\dot{x} < 0$. At the point where $\dot{x} = 0$, the velocity is zero and the fluid stops flowing. Such a point is referred to as a fixed point. It is shown that there are two types of fixed points available: the stable fixed point, represented by a solid black dot, and the unstable fixed point, represented by an open circle. A stable fixed point is also known as an attractor, as the flow nearby heads towards it, while an unstable fixed point is often called a repeller, as the flow nearby moves away from it. In general, the fluid flows from an unstable fixed point towards a stable fixed point in each region. Specifically, any initial starting point that is not a fixed point will follow the direction of the arrows and reach a stable fixed point or approach infinity as time becomes infinite.

Another geometric way of thinking about the fixed point and its stability is to consider the pendulum system shown in Fig. 1.4. Figure 1.4 shows two scenarios: (a) a small-diameter ball hanging from a pendulum boom and (b) a pendulum boom holding such a ball.

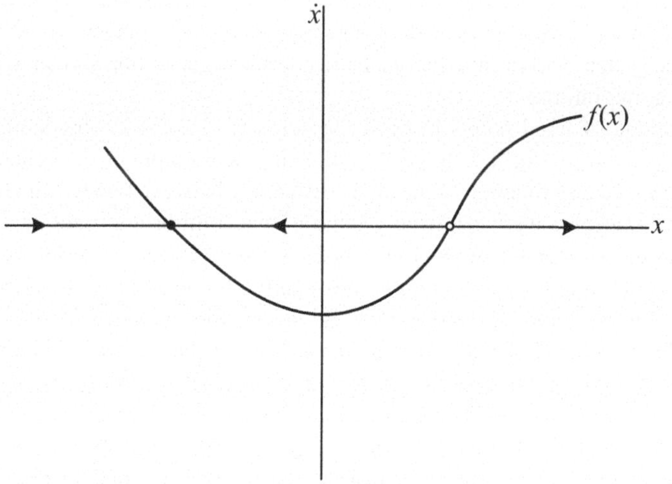

Fig. 1.3 Example of fixed points with their stability

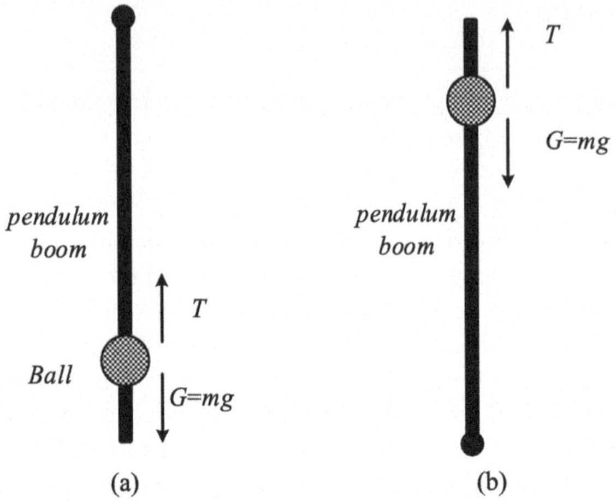

Fig. 1.4 Pendulum system: **a** normal position, **b** upside-down position

At the position shown in Fig. 1.4a, the pendulum boom and the ball achieve an equilibrium condition where the gravity on the ball and the force of the pendulum boom are equal and opposite. This position can be considered to create a fixed point. If a small disturbance is introduced into this system, the pendulum may oscillate for a few cycles, but it will eventually return to the position shown in Fig. 1.4a as duration approaches infinity. Hence, the position in Fig. 1.4a is a stable fixed point. Although the position shown in Fig. 1.4b is also a fixed point, since the balance between the gravity of the pendulum and the force from the pendulum boom to hold the ball is achieved, it is unstable. After the introduction of a small disturbance into the system, the pendulum will swing, but it will never return to the position shown in Fig. 1.4b. Instead, the system will eventually reach the equilibrium position shown in Fig. 1.4a with time going infinite.

An example of fixed points in power systems is to consider the common P_e-δ curve, as is shown in Fig. 1.5. In general, the electrical output of the generator (P_e) can be expressed as a function of the angle difference between two terminals (δ). For a given mechanical power (P_m) from generator, two equilibrium states i.e. two fixed points represented as points A and B, can be found along the curve. Point A is a stable fixed point and attracts the operating points nearby. For example, if a disturbance is injected into the system and pushes the system operating point to A′, the generator will accelerate and the δ value will increase because the given P_m is larger than the P_e produced from the generator. As the time approaches infinity, the system will ultimately settle down to point A. Hence, point A represents a similar scenario shown in Fig. 1.4a. When the system operates at point B, it will either restore back to point A or reach infinite after a small disturbance is injected. Point B in Fig. 1.5 is an unstable fixed point, corresponding to the scenarios shown in Fig. 1.4b.

It should be noted that the term 'fixed point' is used in bifurcation theory to describe mathematical concepts and principles. However, in the context of power

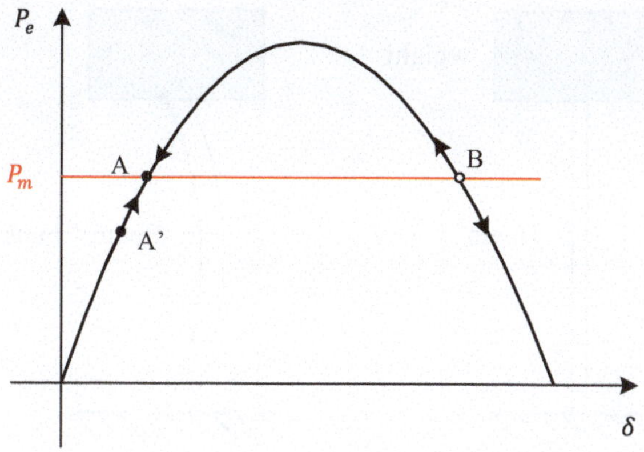

Fig. 1.5 Example of P–δ curve

system studies, a fixed point can be referred to equivalently as an equilibrium point. These two terminologies convey the same meaning in this thesis and can be used changeably.

1.3.3 Various Types of Bifurcation

As discussed in Fig. 1.3, following the vector field of the flows, the solution will either settle down at a fixed point or tend towards $\pm\infty$. However, as system parameters change, the qualitative structure of the flow (i.e. the flowing direction of the fluid) may also change. These qualitative structural changes in the system are typically associated with the emergence and disappearance of fixed points or changes in the stability of fixed points. In general, these qualitative changes in the dynamic vector field are referred to as bifurcations, and the parameter values at which such bifurcations occur are known as bifurcation values [30].

Bifurcations are important because they provide information about the transitions and stability of the system as parameters vary. Figure 1.6 provides a graphic representation of bifurcation by considering the buckling of a beam. If a small weight is placed on the top of the beam, the beam remains vertical and can support the weight, as shown in Fig. 1.6a. However, when the weight is too heavy, as shown in Fig. 1.6b, the beam may buckle, and the system may become unstable. In this example, the weight represents the bifurcation parameter, and the deflection of the beam from the vertical can be considered to represent the system state variable x. As the weight is slowly increased from 0 to ∞, at a certain point the beam will start to bend. This point is the bifurcation point, and the weight that first triggers this bending is the bifurcation value.

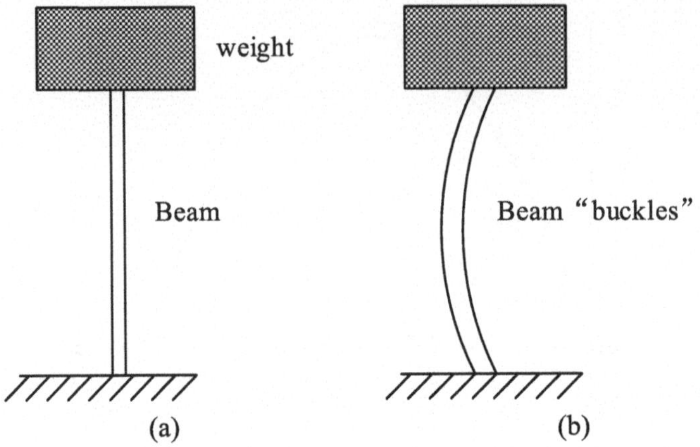

Fig. 1.6 Example of bifurcation: **a** vertical beam, **b** bent beam (at the critical point)

In this thesis, three types of bifurcations, all closely related to instability in power systems, are considered: Saddle-Node bifurcations (SNB), Hopf bifurcations (HB), and Limit-induced bifurcations (LIB).

SNB occurs when two fixed points move towards each other, collide, and then disappear as a parameter varies. A graphic way to understand SNB is to consider the system represented in Fig. 1.7. Suppose the position of the curve $f(x)$ is controlled by a bifurcation parameter u. As u increases, the parabola moves up and two fixed points (one stable fixed point represented by a solid black dot and another unstable fixed point represented by an open circle) move towards each other. At a certain value of u, the curve is tangent to the x-axis. At this point, two fixed points collide and form a half-stable point, which is shown in Fig. 1.7a. Figure 1.7b illustrates that, as the u further increases, no fixed points exists. The state shown in Fig. 1.7a is an example of SNB.

HB is related to the emergence and disappearance of limit cycles and is commonly associated with changes in the stability of fixed points. Figure 1.8 provides a geometric example of HB. In the analysis of power system stability, the system typically experiences damping oscillations after being subjected to disturbances. Assume the decay rate is determined by a bifurcation parameter u. If the decay rate is positive, the magnitude of the oscillation decreases with respect to the time and the system is stable (Fig. 1.8a). As shown in Fig. 1.8c, if the decay rate is negative, the magnitude of the oscillation increases and the system loses its stability. At the critical value of u, the decay rate of oscillation becomes zero and oscillation will continue as the time approaches infinity. This is shown in Fig. 1.8b as an example of HB.

The basic mechanism of LIB is the shift in system equilibrium trajectory from the initial path to a new manifold, as the parameters vary. Consider the system represented in Fig. 1.9a. The parabola shown in Fig. 1.9a is formed by piecewise functions, consisting of parts from different parabolas 1 (dashed line) and 2 (solid

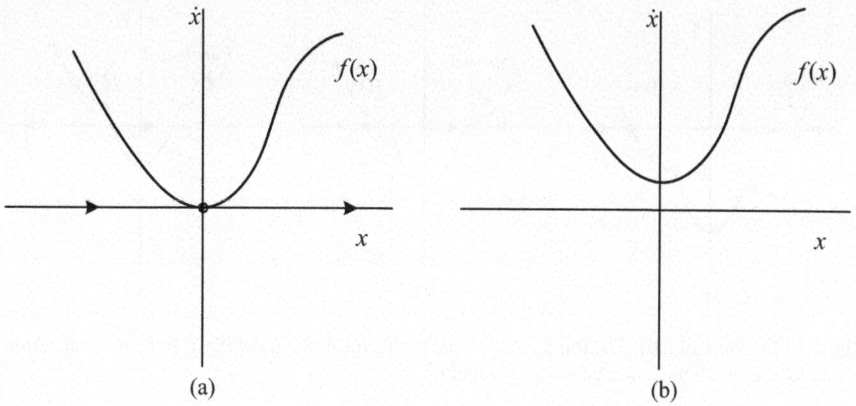

Fig. 1.7 Example of SNB. **a** One fixed point exists, (at the critical point) **b** no fixed point exists

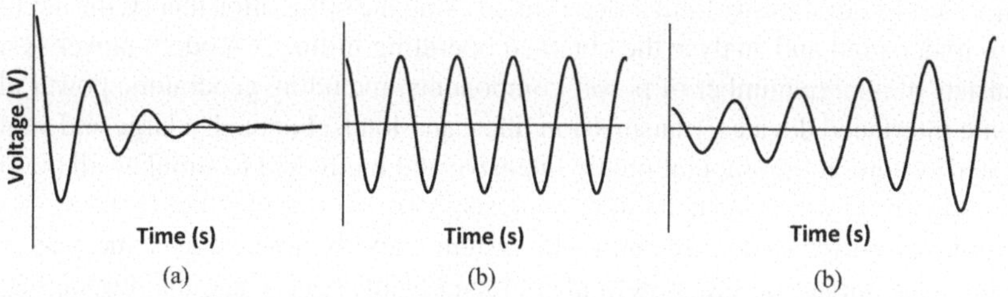

Fig. 1.8 Example of HB. Oscillations with **a** decreasing magnitude, **b** constant magnitude, and **c** increasing magnitude after being subjected to a disturbance

line). Assume a bifurcation parameter u regulates the position of the curve in a similar way that is discussed in the SNB example (i.e. Fig. 1.7). With the increment of u, , the fixed points follow the trajectory of parabola 1 initially (shown in Fig. 1.9a) and then follows the parabola 2 (shown in Fig. 1.9c). At the critical value of u shown in Fig. 1.9b, the trajectory of the fixed points shift from parabola 1 to parabola 2. This critical state is a representation of LIB.

The mathematical conditions used to identify these bifurcations are discussed in more detail in Chap. 2.

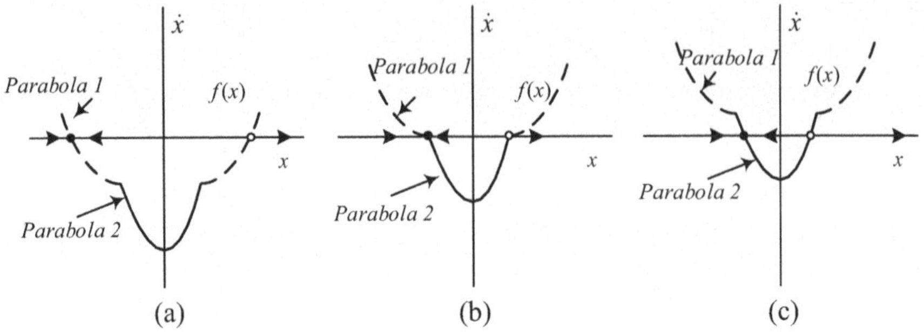

Fig. 1.9 Example of LIB. The trajectory of fixed points **a** follows parabola 1, **b** interchanges from parabola 1 to parabola 2, and **c** follows parabola 2

1.4 Sensitivity Analysis

Once the system stability limit is determined using the bifurcation theory, the next step is to understand and analyse the obtained operating limits. A modern power system consists of a large number of power components, including generators, power electronic interfaced devices, transmission lines, and loads. For such a large and sophisticated system, it is computationally intensive and inefficient to simulate all possible scenarios in power system stability analysis. Moreover, it may not be necessary to include all power system scenarios in system stability studies, as some may have little or no impact on the system phenomena of interest. Thus, this highlights the growing need for sensitivity analysis in power system stability studies.

Sensitivity analysis is an assessment that evaluates how the variations in the input variables impact the corresponding changes in the output variables [32]. The input variables with high sensitivity indices are considered as critical parameters that significantly affect the behaviour of output variables. Sensitivity analysis is an essential part of power system stability assessment. In a typical power system stability analysis, sensitivity studies are performed in order to figure out the critical parameters that have significant impacts on the system dynamics. A thorough sensitivity analysis can further help to determine the optimal parameter setting to keep the system operating away from the stability limits and to guarantee the system's stable operation [33].

1.5 Literature Review

This section provides a review of the existing research related to this thesis, including the investigation of the impact of VSCs on system stability, the application of local bifurcation in analysing power system stability, and sensitivity analysis for power system stability.

1.5.1 *Impact of VSCs on System Stability*

Over the years, abundant investigations into VSC-related power issues have been completed, and various methodologies to improve overall system stability have been proposed. In the past research, various controllers have been designed for VSCs according to a variety of different purposes. The most commonly used control schemes adopted are direct control (power–angle control), vector current control, and power synchronisation control.

This thesis focuses on power issues caused by conventional grid-following vector controls, as these are most widely used in current power systems, as well as being one of the most well-established control schemes in the CIGRE. A review of existing research into VSCs with vector current control, specifically surrounding their effects on power system stability is provided in the following sections.

1.5.1.1 Effect of Single VSC on the System Stability

Stability issues related to single grid-connected VSCs have been investigated thoroughly in previous work. Investigations into this area can be generally classified according to the specific topic concerned as (1) two key factors affecting the stability of power systems with VSC-connected short-circuit ratio (SCR) and phase-locked loop (PLL), (2) interaction between PLL and control loops within one VSC, and (3) interaction between VSCs and the network.

SCR and PLL—Two Key Factors

Most previous work has concluded that PLL and grid stiffness are the two main factors affecting the stability of VSC-connected systems.

The strength of the network is commonly indicated by the SCR index. Generally, a lower SCR value indicates a weaker grid condition. When the SCR is less than 3, the AC network that integrates with the VSC is considered weak [9]. It has been found that VSCs connected to networks with high impedance, which is the main cause of a weak grid with low SCR, are prone to system instability [34–43]. The characteristic roots discussed in [41] clearly demonstrate that the system stability margin is reduced with smaller SCR values, with converter systems becoming more unstable under weak grid conditions. Similar findings have been seen in long-distance power transmission from wind farms with high impedance, which are more likely to suffer from instability, as shown in the case study in [42].

Alongside the effects of weak grid conditions, the work in [42, 44–49] confirms the role of the PLL in system instability within weak grid-connected VSC systems. The instability mechanisms caused by the PLL are investigated in depth in [45, 46] using the developed state-feedback model. It is shown that the loop formed by the PLL continuously drives the system towards instability, and its impact becomes more significant when connected to a weak network [45]. The developed model is also utilised in [46], with analysis expanded to the condition of an islanded power system.

Similar conclusions can be drawn from the work in [47–49] based on the impedance model of the VSC system. The study in [47] attributes the impact of the PLL on system stability as the contribution to the negative impedance and reveals that the range of negative impedance increases if the bandwidth of the PLL increases. Similarly, the results in [49] show that the coupling admittance due to the interaction between the PLL and grid impedance degrades the stable operation of the inverter-grid system when the SCR of the network is small.

Interaction Between PLL and Control Loop

Within a VSC, the speed of the outer feedback loop controls (i.e., the active and reactive power control, the DC voltage control, and the AC voltage control), which operate below the fundamental frequency, is typically ten times slower than that of the inner current control, which operates with a bandwidth above the fundamental frequency [50]. Consequently, the dynamic interaction between these control sets is weak and can usually be considered negligible. However, as shown in [12], the bandwidth of the PLL varies over a wide range, introducing the possibility of interaction with either the outer or the inner loops of the VSC. There has been extensive research into the interaction between control loops within a given VSC and their impacts on the system stability, as in [51–56].

The effect of PLL on the stability of DC link voltage is analysed based on a newly developed small-signal model of a VSC in [52]. It is shown that the PLL may have a negative impact on the DC voltage stability, particularly in a weak grid. The small-signal model in [52] has been further developed in [53] and used to reveal the interactions among the control loops inside the VSC, especially with regard to their impact on the stability limits of an integrated weak power system. The results in [53, 54] clearly show that the interaction between the PLL and DC voltage control constrains system power transfer limits. Similar work can be found in [55], in which a two-degrees-of-freedom mass-spring model is adopted to analyse the coupling effect between the PLL and the DC voltage control. The analytical results reveal that the modal resonance between the PLL and the DC voltage controller reduces the damping of the network. Weak grid conditions will further amplify this damping reduction, causing lower-frequency modes to become unstable. The results in [56] further support this finding based on a sensitivity analysis of different controllers' parameter settings. As is shown in [56], for a very weak grid, a slow PLL bandwidth, combined with a lower bandwidth of the outer loop, is recommended to reduce the mutual coupling between them. Otherwise, the interaction between the PLL and the outer loop would reduce the power transfer limit and further degrade system stability. However, it should be noted that this approach may improve the system's small-signal stability at the expense of reducing its large-signal stability.

Alongside the investigation of the interaction between the PLL and outer loop, several researchers have analysed the interaction between the current inner loop and other controllers [56–58]. The modal analysis results in [57] show that the system may lose its stability due to interaction between the PLL and current inner loop when the connected grid is weak. The interaction between the inner current loop and feedforward voltage controllers is investigated in [58]. The results in [58] not only confirm the positive damping effect of the current control loop on the voltage stability, but also demonstrate that the outer feedforward voltage control forms an additional loop which deteriorates the dynamic performance of terminal voltage.

Interaction Between VSC and Network

Aside from the analysis of the interaction between control loops within a single VSC, some existing research, such as that in [38, 44, 51, 59–75], considers the VSC as a whole component, and investigates the impact of the VSC as a whole on the AC-side network. Among such work, several advanced analytical models of the VSC have been proposed, using a range of analysis techniques.

The work in [44] adopts a conventional state-space model with modal analysis to investigate the dynamic behaviour of VSCs under various external grid conditions. The results show that the grid stiffness and PLL bandwidth are the two key factors affecting VSC system stability [44]. The locus of the critical eigenvalue has clearly demonstrated that a VSC with a high PLL bandwidth under weak grid conditions is prone to system instability.

To investigate the interaction between power electronics and AC systems in power networks, the system is typically divided into two subsystems. In [69–73], power grids are modelled as two subsystems in the form of a VSC-based renewable generation subsystem and a subsystem containing the rest of the system in order to investigate the impact of VSC control design on the full system's small-signal stability. The results in [70] demonstrate that when the open-loop frequency of the VSC-related oscillation mode is close to the open-loop frequency of the electromechanical oscillation of the rest of the power system (known as modal coupling), the system's critical eigenvalue moves towards the right half of the complex plane.

An impedance-based modelling approach, developed from the work in [75], has been widely used to analyse VSC-related power issues, including the interaction between multiple VSCs and the impact of a VSC on network stability. Several impedance-based models with different modelling details are created in [59] to identify the most appropriate VSC model for analysing oscillation issues caused by sub-synchronous interaction between a VSC and a hybrid AC/DC network. The comparative work in [59] is further extended in [60], based on the consideration of different types of outer loops, to offer a comprehensive analysis of the impacts of control schemes and their bandwidths on the impedance modelling of grid-tied VSC systems. A comprehensive and systematic small-signal modelling, analysis, and comparison framework of grid-following inverters is completed in [61]. Both the eigenvalue analysis based on a state-space model and the Nyquist plot based on a dq-domain admittance model are employed in [61] to analyse the stability of a VSC system. The results in [61] clearly demonstrate that both methods can be used to perform small-signal stability analysis of power systems with connected VSCs.

The Bode plot is another effective analysis technique for evaluating VSC performance based on the impedance-based model or transfer function [38, 51, 62–67]. Based on the Bode plot of a developed VSC impedance model, the sub-synchronous interaction between the inverter and a nearby synchronous generator has been studied [51, 66].

Other research has applied nonlinear analysis approaches, such as bifurcation-based analysis, to investigate how a VSC affects both the small-disturbance stability of the system [62–64] and the voltage stability [67, 68]. VSC-related oscillatory instability appears and is identified as HB in [63]. The stability boundary formed by HBs has been built based on numerical studies and verified using time-domain simulation. The study has clearly illustrated that bifurcation theory can be used to accurately perform the system stability assessment.

1.5.1.2 Effect of Multiple VSCs on the System Stability

Alongside the past work on single VSC-related power issues discussed in the previous section, research into the interaction between VSCs and the impact of such interaction on system stability has also been conducted. Based on the different types of coupling networks used, the past work on the interaction among VSCs can be divided into three main categories: (1) interactions coupled through the AC network, (2) interactions coupled through VSC-HVDC links, and (3) interactions coupled through a multiple-terminal DC network (MTDC) (i.e. the DC network formed by connecting multiple VSC-HVDC links).

Interactions Coupled via the AC Network

Much previous research has investigated the interactions between VSCs coupled through the connected AC network and analysed the impact of such interaction on system dynamics [76–85]. A completed state-space model of an inverter-based microgrid consisting of multiple VSCs is built in [76] to facilitate interaction studies. The dynamics of the VSC, network, and load are represented by differential equations included in the individual sub-modules, and all sub-modules are then combined using a common reference frame to form the overall system model. The calculated eigenvalue, combined with eigenvalue sensitivity analysis, is then used to investigate the interaction between the VSC controllers and the interaction among the inverters and the other network dynamics. A relative gain array (RGA) based on the developed multiple-input-multiple-output (MIMO) state-space model is applied in [77] to identify interaction among the controllers of different VSCs in the power plant. The modal analysis, combined with RGA, has clearly demonstrated the existence of interaction between VSCs if they are connected closely (i.e., the electrical distance between their point of common couplings (PCCs) is short). As is shown in [77], the adverse interaction is associated with the selection of the type of VSC control loop used and improper parametrisation of the controllers. The study in [78] further confirms these findings. The eigenvalue locus combined with the participation factor analysis from [78] shows that a low active power flow and a long AC line generally help reduce the coupling effect between VSCs.

Apart from the modal analysis, the Nyquist plot of an impedance-based model is proposed in [79], and this is now widely utilised in the analysis of interaction among VSCs. An impedance-based approach is used in [80] to investigate the effect of the interaction among power electronics coupled via AC networks on the system stability. Similar to work in [39, 81], the work in [80] focuses on the emergence of unstable low-frequency oscillation, which is represented as HB based on bifurcation theory. The results clearly show that the stability boundary formed by HBs shrinks when multiple VSCs are integrated into the AC network. Specifically, the mutual interaction among VSCs makes the system more prone to instability. This emphasises the necessity to consider the interaction among VSCs via the AC networks when analysing system stability. The analysis based on an extended impedance model shows that the interaction between the dynamics of multiple converters is heavily dependent on the control parameter configurations and the tuning of the HVDC lines [82].

In addition to the impedance-based approach, some studies, such as those in [84, 85], have analysed system stability based on the developed transfer function of the VSC system. The impact of the interaction among VSCs via the AC network on the DC voltage stability is assessed based on a simplified VSC transfer function in [84]. With the developed transfer function model, interactions between VSCs are studied with the consideration of different grid strengths, the operating point, and controller bandwidth [84].

Although much effort has been made to investigate the interaction among the outer loops of VSCs where the frequency range of interest is below the fundamental frequency, interactions above the synchronous frequency could also exist between VSCs, especially between the inner current loops of coupled VSCs [86–88]. The interaction among VSCs, focusing on the inner current control time scale, is also examined in [87] based on a simplified state space model. The study in [87] shows that the grid strength and current controller bandwidth have significant impacts on the dynamics of interaction among the inner current loops of different VSCs. Apart from interactions among control loops of different VSCs, the interaction among VSCs could also occur due to frequency coupling between the switching frequency and the inner current loop frequency. It is shown in [88] that the harmonic distortion introduced by the network may have a similar frequency bandwidth to the inner current loop, resulting in system instability.

Interaction Coupled via VSC-HVDC Links

To construct a VSC-HVDC link, two VSCs are connected via an HVDC link, which naturally indicates potential interactions between the VSCs. Besides, two networks, the AC and DC networks, are integrated through the VSC-HVDC link. Hence, there may exist potential dynamic interactions between the DC and AC networks. For a single VSC-HVDC link, previous interaction work can be classified into three categories: (1) the interaction among VSCs via the DC network (i.e. via HVDC) [89–91], (2) the DC/AC interaction between the VSCs and the AC/DC network [92–94], and (3) the interaction between the VSC-HVDC and the connected AC network [95–99].

In a typical VSC-HVDC link, it can be assumed that one VSC uses the DC voltage outer control loop and another VSC adopts the active power outer controller. The work in [90] analyses the interaction among the VSCs via an HVDC link using an impedance-based model. The full VSC-HVDC link is divided into two sub-systems.

The output impedance of the source sub-system (that includes VS_C with DC voltage control) and the input admittance of the load sub-system (that includes VSC with active power control) are derived. A Bode plot of the ratio of source and load impedance is used to assess the system phase and gain margins, which offer key information about the system's stability. A similar impedance model is adopted in [91] to calculate the system phase margin, considering interaction among converters via the DC network in a VSC-HVDC link.

Complex torque analysis is applied to a closed-loop model to assess the small-signal stability with respect to the interaction between DC/AC networks [94]. These results clearly demonstrate that the DC network loading, the length of the DC line, and the bandwidth of the DC voltage controller in the VSC are the critical parameters affecting the interaction dynamics between the VSC and the associated AC grid.

In addition to many studies focusing on the interaction within AC/DC networks coupled through VSC, there have been studies concerning the impact of the HVDC link on overall system dynamics [95–99]. A generic RMS model for MMC-VSC-HVDC has been developed based on the averaged EMT model to investigate its impact on the system's dynamic response in accordance with transient stability and small-signal stability [95]. The results show that the linearised generic MMC model can be used to analyse the influence of the HVDC link on the interconnected large AC system. A thorough assessment of the impact of VSC-HVDC on the interdependency of different types of stability (i.e. small-disturbance rotor-angle stability, large-disturbance rotor-angle stability, frequency stability, and voltage stability) is completed in [97]. The Spearman correlation is adopted to quantify the influence of the HVDC link. Based on the simulation results, it is apparent that the inclusion of additional HVDC helps to support power system operation and increases the positive correlation between various types of stability. The investigation has also demonstrated that the system loading, HVDC controller, and grid topology have significant impacts on the strength of interdependency between stability phenomena.

Interaction Coupled via MTDC

Unlike LCC-HVDC, which requires a complex configuration, it is relatively easy to construct a multi-terminal system using a VSC-HVDC with constant voltage polarity. An MTDC system is formed by connecting multiple VSC-HVDCs [100]. Some previous work has focused on analysing potential interactions between various numbers of VSCs in such an MTDC network [101–105].

The conventional state-space model combined with modal analysis is a typical technique for investigating the coupling effect between converters in MTDC networks [106]. The work in [105] analyses interactions between VSCs in an MTDC network and system stability margins using a mode identification technique based on the participation factor analysis and the eigenvalue sensitivity. The study clearly reveals the existence of interactions among VSCs coupled via MTDC. It is shown in the results that when the PLL gain increases or the SCR value of the interconnected network is lowered, the coupling effect between converters becomes stronger, and the stability of the MTDC may degrade. Similar work can be found in [101], where a stepwise eigenvalue analysis is adopted to identify the interaction between VSCs in the MTDC and to analyse the impact of these interactions on system stability as system parameters and operating points vary. The work in [101] is further extended in [102] to develop a classification method to distinguish the local converter mode primarily related to the one-terminal DC voltage and interaction mode, which consists of multiple terminals of various DC sub-systems. A transfer-function-based approach can also be used to investigate interactions among VSCs in an MTDC network. In [103], the system is divided into two small open-loop sub-systems, one consisting of selected VSCs and another containing the remainder of the MTDC network. This work indicates that when the MTDC implements voltage droop control, it may suffer from an open-loop modal coupling (i.e., between the open-loop oscillation of the selected VSC and the open-loop oscillation of the rest of the MTDC grid), which can cause the system to lose stability. The work in [104] further studies the occurrence of this open-loop modal coupling between AC and MTDC systems.

1.5.1.3 Summary

Previous work in this area has typically focused on revealing and investigating the instability mechanisms triggered by interaction among VSCs, and there is a lack of methodology to identify system stability limits efficiently.

Additionally, most existing sensitivity analyses within VSC-related stability studies have been predominantly tested within a small varying range and only a limited number of scenarios are considered. Hence, the impact of parameter variations on the system stability over a wide range, has not yet been fully examined.

Moreover, as existing studies into VSC-related issues have been limited to a small range of operation points, systematic investigations of controller performance across wide-ranging conditions considering interactions between converters have not been completed yet.

Additionally, most past work has focused on a single type of stability when performing stability assessments, despite the fact that various types of stability are correlated, particularly when power systems with multiple power electronics are connected. Although a few studies have analysed system stability across multiple types of stability, they typically analyse each classification of instability individually. Therefore, no thorough assessment of system stability considering multiple types of stability exists.

To capture the newly emergent power electronics-related issues exemplified by interaction among VSCs, some of the past work—for example, the work in [107, 108]—has analysed system stability by modelling the AC network using sets of differential equations. However, dynamic modelling of a full system with differential equations would result in a prohibitively large number of state variables being included and, consequently, require a prohibitively long simulation time.

1.5.2 *Application of Local Bifurcation Theory in Power System Stability Analysis*

As discussed in Sect. 1.3, bifurcation theory is used to identify qualitative changes in the topological structure of a system. In power systems, structural change is commonly related to changes in system stability. Bifurcation theory thereby provides a natural platform for studying power system stability, as it allows different power system issues to be represented as various types of bifurcations. The next few sections provide an overview of the application of local bifurcation theory in previous system stability studies across different classifications of stability.

1.5.2.1 Voltage Stability

For a typical power system, an increment in the system's active power loading will result in a drop in the bus voltage. This is revealed by the relationship between the active power demand and bus bar voltage, known as the P–V or nose curve. In conventional voltage stability analysis, the tip of the nose curve represents the system's operational limit, which is the maximum power transfer that the system can handle. In bifurcation theory, the nose curve can be viewed as a set of fixed points in the system. The tip of the curve is represented as an SNB, which is a typical static bifurcation point.

Early research into voltage stability primarily focused on the maximum power transfer points represented by SNBs. Hence, conventional system voltage stability analysis based on SNBs has been thoroughly studied [109, 110]. The loading margin between the given operating point and the SNB is typically used as an index for loadability assessment. The work in [110] adopts a direct method to compute the SNB and uses the loading margin as an index to assess system stability under various contingencies. Further studies in [111] indicate that a system is prone to instability if the reactive power limit of the synchronous generator is considered. Based on this, the work in [112] shows that the system may immediately change from a stable to an unstable condition when a reactive power limit is encountered. In such cases, it is the reactive power limit rather than the SNB that defines a system's maximum power transfer point (known as the loadability point). This phenomenon is typically referred to as the LIB in the literature. The SNB and LIB are commonly classified as static bifurcations, as they are associated with the emergence and disappearance of system operational points. Beyond these static bifurcation points, there is no equilibrium point available for the system to operate at. Specifically, no operational points exist, as the system loading increases exceeding the static bifurcation and leading to the voltage collapse.

However, not all system outages are caused by static bifurcation points; some voltage stability issues are related to dynamic system events. The work in [113] demonstrates that a dynamic bifurcation point—for instance, an HB—can also cause system instability. An HB is a typical dynamic bifurcation, which can be identified using modal analysis. For a given initial operating point, the system may reach an HB and become oscillatory unstable.

Different Analysis Methods

The following section reviews the various analysis methods used to identify bifurcations in power systems. This includes (1) direct method [114–117], (2) continuation method [116, 118–121], (3) optimisation-based method [122–127], and (4) machine-learning-based approach [128, 129].

A power system is a typical nonlinear dynamic system from a mathematical modelling perspective. Based on this, power systems are commonly modelled as sets of differential equations to capture their dynamic characteristics. The direct method, which is widely used to identify the point of voltage collapse in the conventional voltage stability study, can be used to compute the bifurcation points with their mathematical condition [114]. This method is employed in [115–117] to identify the SNB along the system equilibrium curve, with the overall system loading considered as the bifurcation parameter. Although the system equilibrium trajectory can be obtained by continuously solving the mathematical modelling of a power system with different system loading values, it will encounter convergence issues, particularly when the system approaches an SNB, where the system matrix becomes singular.

To overcome this shortcoming, the continuation method is proposed. In general, the continuation method consists of two steps: the predictor, which anticipates the next operational point based on the current system equilibrium solution; and the corrector, which identifies the real fixed point on the equilibrium trajectory based on the initial approximation obtained from the predictor [118, 119]. Therefore, the continuation method has been adopted in power system analysis to trace the system equilibrium curve [120, 121]. The work in [116] compares the results using the direct method and the continuation method. It can be concluded that the continuation method can be used to compute the bifurcation point as accurately as the direct method does. Compared to the direct method, the continuation method provides more information regarding system behaviour as bifurcation parameters vary. In terms of its application in power systems, however, the continuation method is computationally expensive, particularly for large power systems [116].

Alternative ways to calculate the bifurcation include the optimisation-based approach, which can be viewed as an updated version of the direct method. The differential equations that represent the dynamics of a power system in the direct method are reformulated as an optimisation problem [111]. Consequently, the voltage stability problem is stated as an optimisation problem, which allows the use of optimisation algorithms such as Interior Point Methods [122, 123], Sequential Quadratic Programming [124], and Genetic Algorithms [125, 126], to identify the bifurcation points. Similar to the direct method, the optimisation-based method is computationally efficient but it is less informative and reliable than the continuation method. A comparative work examining system voltage stability using the continuation method and optimisation-based approach was completed in [127]. The results clearly demonstrate that both methods can be employed to compute system operation limits.

Alternative solutions for evaluating the maximum loadability of a system include the use of machine learning techniques [128, 129]. The machine learning method can be viewed as a combination of both the continuation method and the optimisation-based approach. The work in [129] is a typical example, whereby the continuation method is initially adopted to generate a certain number of system loadability points that form the system stability boundary. A neural network representation of the obtained stability boundary of the system is then generated using back-propagation neural network (BPNN) training. Following this, a differentiable mapping function that approximates the stability boundary of the system is extracted from the BPNN and is then incorporated into the optimisation formula as a constraint to allow further system loadability studies to be performed. It should be noted that the issues discussed within this thesis are convex. Therefore, the review in this section focuses on optimisation-based methods for convex problems. The non-convex problem is out of the scope of this thesis. A comprehensive review of various optimisation algorithms adopted in power system analysis for both convex and non-convex problems can be found in [130–134].

Stability Analysis Considering Multiple Bifurcations

The work in [112] demonstrates that the system may encounter an LIB before reaching the SNB as the loading on the system increases. The limits considered in LIBs are typically the reactive (Q) limit or the generator field voltage (E_{fd}) limit. Once these limits are encountered, the system may either become unstable immediately or operate following a new equilibrium trajectory as the loading increases. Apart from the uncertain situation after reaching these limits, the introduction of the LIB also causes the previously smooth equilibrium curve to become unsmooth, causing further challenges for the identification of the operational limits of the system. Much of the previous research has focused on incorporating LIBs when analysing the system's voltage stability. The continuation method is found to be advantageous in incorporating LIBs into the system loadability analysis [135], but it may suffer from convergence problems for certain LIBs that significantly change the solution manifold.

To address this, the work in [136, 137] develops a new continuation method to calculate system bifurcations along the system equilibrium trajectory, taking both the LIB and SNB into account simultaneously. The studies demonstrate that all operational limits can be effectively identified with respect to a continuous change in the system loading until a sudden instability or an SNB occurs. A similar, modified continuation method is proposed in [138] to compute SNBs and LIBs in a power system. The resulting quantitative comparison shows that the proposed method can produce results accurately with reduced execution time. Alternatively, the optimisation-based approach is used. The work in [139, 140] presents an optimal power flow with complementarity constraints that model the behavioural changes due to the reactive power limits of the generator. The mathematical program developed in [139, 140] is further developed in [127] to perform a comprehensive and detailed study of system voltage stability that considers both the SNB and LIB simultaneously.

Although techniques mentioned above have been developed to introduce LIBs in identifying bifurcations in power systems, these applications have been limited to power flow calculations, particularly in power system models where the dynamics of power components are neglected. If the dynamics of power components are included and represented by sets of differential equations, it is necessary to consider generator excitation limits when performing voltage stability assessments. This includes modelling the generator dynamic response before and after generator field limits as different sets of differential equations and including them as two inequality constraints in the optimisation problem formulation. An additional equality constraint is also introduced into the optimisation problem formulation to ensure that at least one of the equality constraints that represent the observed behavioural changes due to exciter limits holds [141–143]. Based on such an optimisation-based framework, [142] shows that the system loadability surface is non-smooth and consists of multiple equilibrium manifolds intersecting at corner points. An alternative solution to include the E_{fd} limit during the voltage stability assessment is to use differential equations to approximate the changes in the dynamic behaviour of the system due to limits [144]. Consequently, the previous non-smooth system equilibrium trajectory becomes smooth again. Hence, the conventional continuation method and optimisation algorithm can be employed to analyse system voltage stability with the simultaneous consideration of the SNB and LIB.

When the bifurcation parameter is multi-dimensional (i.e., there is more than one parameter driving the system to bifurcation), a system loadability boundary formed by bifurcations can be obtained. For such a system loading boundary, it is of interest to identify the bifurcation locally closest to the current operating point. The Euclidean distance in the parameter space to this bifurcation point represents the worst case (i.e., the shortest loading margin). A methodology to compute the closest bifurcation on the hypersurface formed by SNBs to the current operating point is proposed in [145]. It has been clearly demonstrated in [145] that the vector pointing from the current operating point towards the locally closest SNB is normal to the system stability hypersurface formed by SNBs. Hence, the system's shortest loading margin can be computed using the obtained normal vector. Moreover, the normal vector also indicates the sensitivity of this closest bifurcation with respect to other parameters. In [146], this shortest loading margin is then formulated as an optimisation problem to determine the optimal shunt and series compensation parameters to maximise the distance to the locally closest SNB. The shortest loading margin is maximised in the parameter space using an optimisation approach in [147] that considers uncertainty in the system loading. In general, previous work has formally and systematically investigated the system's maximum loadability points by considering the SNB or LIB [128–135].

The work in [148] analyses the system loadability curve taking both SNB and HB into consideration simultaneously. The results clearly show that as the loading increases, the system may reach an HB before reaching an SNB. Based on this, several papers have discussed the impact of the HB on the system loadability boundary by considering voltage stability and propose various optimisation algorithms to improve these [149–157].

The effect of generator exciter control on the occurrence of the HB was examined in [149] based on a simplified linearised power system model. The work in [150] employs a generation re-dispatch technique to eliminate the HB in power systems considering uncertainty in the loading and power distribution. Several previous researchers have managed to develop indices that can predict the oscillatory instability associated with the HB based on eigenvalues, singular values of states, and the extended system matrix [151, 152]. Similar to the work in [145], a new method for estimating the locally closest HB given a current operating point is proposed in [153]. Rather than adopting the iterative method used in [153, 154], the work in [155, 156] uses optimisation techniques to compute the shortest distance between the current operating point and the hypersurface formed by HBs. This shortest HB margin is further investigated in [157] across different levels of uncertainty in the system demand and the renewable power generation.

A detailed bifurcation analysis of a real multi-parameter power system is completed in [158–162], where the system stability boundary is computed by calculating the system equilibrium trajectories with different parameter configurations using the continuation method. The bifurcation diagrams clearly illustrate that when multiple bifurcations are considered simultaneously, the system loadability boundary becomes a mix of different types of bifurcations.

1.5.2.2 Small-Disturbance Stability

Apart from voltage stability, local bifurcation can be employed to analyse other types of stability, including converter-driven stability, resonance stability, and small-disturbance rotor-angle stability. As discussed in Sect. 1.1, such stability is referred to as small-disturbance stability and can be analysed using modal analysis. In applications of bifurcation theory, the instability in previous small-disturbance stability studies has thus commonly been represented as HBs.

A general optimisation problem formulation based on eigenvalue calculation is proposed in [125, 126] to identify HBs and their corresponding small-signal stability boundaries in multiple system parameter spaces. Based on the proposed optimisation problem formulation, the minimum distance between the current operating point and boundary can be computed, which can be further used to identify the optimal parameter setting for security control. This optimisation-based method is further extended in [163, 164], with two time-efficient computation algorithms to determine the small-signal stability boundary formed by HBs. Both algorithms are based on the Newton type and show good convergence in the neighbourhood of the solution.

As discussed in the previous section, bifurcation theory has been widely used in analysing VSC-related issues in [62–64, 80, 165]. The dynamic performance of three-phase VSCs connected to non-ideal power systems is studied in [63]. Based on modal analysis, the low-frequency instability due to VSCs, which appears as an HB phenomenon, is observed. The system stability boundary limited by HBs has been identified using numerical studies and verified by means of time-domain simulation.

Similar work can be found in [39], where the low-frequency bifurcation of weak-grid-tied VSC considering PLL is analysed based on a linearised state-space model. The results suggest that the strength of the grid and the active power flow are the two key factors affecting the occurrence of HB. In addition to the analysis of system stability with a single VSC connected, bifurcation has been adopted to analyse the coupling effect between VSCs in [80, 165]. Similar to stability studies when one VSC is connected [63], when based on bifurcation theory, power system instability due to interactions among converters appears as HBs [80]. The system stability limits are then identified using the Nyquist stability criterion of the developed impedance model, and the corresponding boundaries formed by the HBs can be derived in various parameter planes using a cycle-by-cycle simulation.

1.5.2.3 Summary

Although system voltage stability has been thoroughly investigated using bifurcation theory, previous work has tended to analyse system loadability with the consideration of the SNB and LIB and with the HB separately, despite all bifurcations being strongly related to each other. There is a lack of work on power system stability considering all three bifurcations, SNB, LIB, and HB simultaneously.

Additionally, the optimisation methods developed in previous research are based on derived mathematical equations to compute the system's loadability boundary by considering each type of bifurcation separately. No studies to date have developed optimisation-based frameworks to identify and analyse the system's loadability boundary while considering SNB, LIB, and HB simultaneously.

1.5.3 Power System Sensitivity Studies

As is discussed in Sect. 1.4, the aim of sensitivity analysis is to determine the most influential factors affecting the dynamic response of the system based on variations in the parameters considered. The application of sensitivity analysis approaches in power system research has been reported in areas such as power component ranking, correlation assessment, and the identification of system critical parameters. A number of techniques have been applied in the past work. These can be generally classified as (1) analytical method and (2) numerical method, which are discussed in the following sections.

1.5.3.1 Analytical Method

The analytical sensitivity technique is based on a detailed mathematical representation of the power system, which characterises the dynamic response of the system. The eigenvalue analysis based on the analytical sensitivity technique has been widely adopted in power system small-disturbance stability studies [10, 105, 124, 166–170]. The eigenvalue sensitivity technique proposed in [10, 166] provides information about the expected variations in the concerned eigenvalue given a change in a specified parameter. Eigenvalue sensitivity has been implemented in [167] with respect to the generation rescheduling to suppress power oscillations. The calculated sensitivities are used to determine the most effective rescheduling to improve the damping of the concerned mode. Apart from its application in the ranking of power devices, eigenvalue sensitivity can also be used to identify the critical mode that first crosses the imaginary axis as the selected parameter varies [168]. Once the critical mode is determined, the eigenvalue sensitivity can further be used to compute the stability boundary of the system by either applying a test function or using a more direct method [124]. In that work, the eigenvalue sensitivity with respect to state variables is calculated at each iteration, providing the search direction for the next iteration. Similarly, eigenvalue sensitivity is adopted in [169, 170] to trace the small-disturbance stability boundary of the system. The work in [105] develops an analytical framework that combines the participation factor and eigenvalue sensitivity indices to identify and analyse the mode that best represents the interaction between VSCs via the MTDC network. It is clearly shown that the degree of controller interaction can be quantified using the developed sensitivity coefficients. Besides, eigenvalue sensitivity is further used in the calculation of the stability margin of the system, as verified by the eigenvalue trajectory.

In addition to eigenvalue sensitivity, a sensitivity analysis technique combined with the bifurcation-based approach can be used to improve the overall stability of the system [171–173]. The work in [171] derives linear and quadratic estimations of variations in the loading margins of the system with respect to the individual system parameter. The developed sensitivity function can be used to quickly produce an estimation of the change in loading margin given a particular change in the system parameter of interest. Consequently, the proposed sensitivity derivation function in [171] is further adopted in [172] to avoid voltage collapse caused by the SNB. Based on the loading margin sensitivity function, a method is proposed in [172] to calculate the optimum direction in the control space required to maximise the loading margin. The sensitivity of HBs with respect to system parameters is derived in [173]. Similar to the work in [171], the HB sensitivity is easily obtained by computing the normal vector to the HB hypersurface in the parameter space. The numerical results indicate that the calculated sensitivity indices can be used to optimise the loading margin to the system hypersurface formed by HBs.

1.5.3.2 Numerical Method

Unlike analytical methods, which directly calculate the sensitivity based on a derived mathematical model, numerical methods determine the relationships between variables based on a computed dataset. Initially, the method involves generating sampling points based on the specific requirements of the sensitivity analysis technique to be used. A Monte Carlo simulation of each pre-selected sampling point is then performed, and for each computational scenario, the standard stability assessment is completed. Following this, various numerical sensitivity techniques are employed to compute the sensitivity between parameters based on the selected scenario.

Some existing work has calculated eigenvalue sensitivity based on numerical methods. For example, a numerical method-based eigenvalue sensitivity analysis has been used to rank uncertain parameters based on their impact on system modes of concern to reduce the number of uncertainties considered [174]. To achieve this, each uncertainty parameter of interest is varied over a small interval around its nominal value. The corresponding changes in the eigenvalues are obtained, and the sensitivity measure for each eigenvalue with respect to the individual parameter can be calculated.

A thorough comparative assessment of nine sensitivity analysis techniques in terms of their ability to prioritise uncertain parameters affecting system small-disturbance stability is completed in [175]. Within that work, the analytical sensitivity is referred to as the local sensitivity technique as it neglects the high-order cross-effect between parameters. Unlike local sensitivity analysis methods, global sensitivity analysis techniques can numerically investigate the relationships among parameters by considering parameter interdependency over the entire parameter range [175–180]. The results in [175] clearly demonstrate that local sensitivity analysis techniques fail to determine the same influential parameters as those from global sensitivity techniques. This indicates that the application of local sensitivity methods may result in potentially misleading results when considering large complex power systems. An example of a commonly used global sensitivity technique is multi-attribute decision-making (MADM), which is used to investigate the preference of market participants in terms of evaluating the integration of renewable generations into the Australian National Electricity Market [176]. MADM performs sensitivity analysis based on a decision matrix that considers scenarios that are equidistantly sampled across the search range. The Technique for Order Performance by Similarity to Ideal Solution (TOPSIS) algorithm is then employed to investigate the built decision matrix and to rank the preferences of different market participants.

The techniques discussed in [175–179] all require specific sampling points across the search space; for example, MADM generates samples that are equidistant in the range of each parameter; the Morris screening method follows a multi-dimensional search trajectory, and samples for the Sobol method are determined in a symmetrical geometric orientation. Another promising approach is the correlation coefficient reviewed in [180]. The Spearman rank correlation coefficient is used to analyse and quantify interdependency between various types of stability. Unlike the other global sensitivity techniques mentioned above, the Spearman correlation does not have any specific sampling requirements, thus allowing direct determination of the dependency among parameters based on the randomly generated sample. It can be concluded from the work in [175] that the correlation coefficient can produce identical sensitivity results to those produced by other global sensitivity analysis techniques, such as the Sobol and Borgonovo methods, but with a significantly reduced computational effort.

1.5.3.3 Summary

The analytical sensitivity method is reliant on the accurate modelling of power system dynamics, particularly the derived mathematical model of the power system. When multiple types of bifurcations are considered, it is hard to derive a generalised equation to represent the relevant stability boundary. Under these circumstances, the previously developed analytical method is no longer applicable.

Although global sensitivity has been adopted in some previous research to rank parameters based on their impact on the system stability, such investigations have been performed within small parameter search ranges. This neglects the impact of different grid strengths which have been proven as a vital factor affecting system stability. Additionally, most previous sensitivity analyses have focused on the effect of variations in critical parameters on one specific type of stability at a time. There is no analysis framework to identify the most influential variables affecting system stability in large power systems over a wide operating range whilst considering the effect on multiple types of stability simultaneously.

1.5.4 Summary of Past Research

Having reviewed the existing work in the field, several areas identified still require further investigation. A summary of the research gaps thus identified is offered below:

1. A thorough assessment of system loadability/stability limits based on simultaneous consideration of multiple types of stability using bifurcation theory should be completed. This is required as previous research has typically analysed system operation limits by considering each type of stability individually.
2. There is a need to develop a new optimisation-based method to improve the overall loadability of a system considering multiple bifurcations simultaneously. To date, studies have completed several thorough investigations of system loadability using optimisation techniques based on bifurcation theory; however, these are based on considering each type of bifurcation individually.
3. A methodology that can efficiently identify the feasibility region boundary of a system in the multi-dimensional space is currently lacking, as existing research into system operating limits has focused on revealing the instability mechanisms underlying power system phenomena, with the investigation around the system stability boundary limited to a low-dimensional space.
4. There is a requirement for a sensitivity methodology to identify the most influential parameters affecting system stability by considering multiple types of stability simultaneously. Although few sensitivity studies have investigated the impact of parameter variations on multiple types of power system stability, they typically examine each classification of stability individually.
5. A comprehensive assessment of the performance of various VSC controllers across wide operating scenarios is required. Although thorough studies have

been conducted into the impact of VSCs on the dynamic response of a system in the past, there has been limited research on evaluating the performance of VSC controllers across a very small set of operating conditions and the impact of the grid strength on the performance of different controllers is not thoroughly evaluated.

6. There is a lack of a network modelling methodology that efficiently captures newly emergent issues caused by the interactions between VSCs and the network in large power systems. Existing research into these issues has typically been carried out using small-sized power systems with external AC networks represented by Thevenin equivalent circuits. However, such simplified network representation fails to accurately represent the dynamic response of VSC-related issues in large systems. Although a few researchers have analysed the dynamic characteristics of VSCs in large power systems, it is crucial to incorporate the dynamics of such networks into models to capture high-frequency oscillations that result from the interaction between converters and the network. However, this, in turn, generates a heavy computational burden.

1.6 Research Aim and Objectives

This thesis attempts to address the research gaps identified within the field of research as discussed in the literature review section. The main aim of this study is to use bifurcation theory to analyse a wide range of power system instability, particularly with respect to VSC-related power system issues, as well as adopting a bifurcation theory-based approach to efficiently identify system stability limits taking into account various types of stability simultaneously.

To achieve this aim, the following research objectives have been identified:

1. To summarise and critically evaluate the existing literature on both the newly emergent power system issues caused by the VSC and the application of bifurcation theory in power system stability analysis to establish a general picture of the topic.
2. To thoroughly investigate the loadability and stability limits of the system considering several different types of stability simultaneously by applying bifurcation theory-based approaches.
3. To propose a methodology to efficiently identify system stability limits and boundaries in a multiple-dimensional space where several parameters may vary simultaneously.
4. To develop a framework to optimise overall system stability based on the updated loadability concept, with different types of instability represented as bifurcations in the given parameter space.
5. To develop an analysis framework to numerically and accurately identify the most influential parameters affecting system stability considering the inter-dependency among parameters and sensitivity over the full parameter range.

6. To evaluate the dynamic performance of various types of VSC controllers (i.e. the PQ and PV_{ac}) over a wide range of loadability scenarios and different grid strengths to identify the most appropriate control schemes to maintain the system's stable operation.
7. To develop a system modelling approach that captures the dynamic characteristics of critical power system components and simplifies the dynamics of the rest of the network by offering a simplified representation to more efficiently calculate system loadability.

1.7 Main Contributions of the Thesis

The work within this thesis contributes to a number of areas of power system research, specifically regarding the stability analysis of power systems with multiple VSCs connected to the AC side of the network using bifurcation theory. The main outcome of this thesis is the comprehensive joint assessment of multiple types of power system stability (i.e., the simultaneous consideration of both small-disturbance and voltage stability) with the application of bifurcation theory. The bifurcation theory-based analysis in this research can be used to identify and analyse various power system issues, particularly newly emerging issues related to power electronics.

The main contributions of this thesis can be summarised as follows:

- An optimisation technique has been developed to efficiently identify the new loadability boundary that consists of various power system operation limits represented by multiple bifurcation points. This technique can be used to improve the system's shortest loading margin to the new loadability boundary from any given operating point within the control parameter space.
- The development of a methodology to efficiently identify the stability limits concerning multiple types of stability where multiple parameters may vary simultaneously. Compared with previous work in this area, where the research is limited to a two-dimensional parameter space, the use of quasi-random sampling results in the systematic identification of the stability boundary of the system in a multiple-dimensional space with the consideration of the interdependency among parameters.
- An analysis framework is proposed to investigate the impact of key parameters under various grid scenarios on system stability. The partial Spearman correlation coefficient is employed to quantitatively assess the impact of parameters on the dynamic behaviour of the system considering multiple types of stability. The proposed methodology can provide more accurate sensitivity assessments than before with the consideration of global sensitivity over the entire parameter range.
- An analysis framework is developed, based on bifurcation theory, to identify system loadability limits considering both small-disturbance and small-disturbance voltage stability. The necessity of considering both types of stability when calculating the loadability of the system is illustrated by results obtained from multiple test networks. The formulation of the loadability limit of the system

is updated to include concurrent consideration of small-disturbance stability and voltage stability.

- VSC-related stability analysis previously limited to a small operating range is now revealed and analysed over a wide range of scenarios by the application of bifurcation theory. The work shows that bifurcation theory can effectively identify and analyse the instability caused by interactions between VSCs and the AC network in both large and small power systems. A thorough investigation into the effects of different VSC control schemes on the system loadability over wide-ranging scenarios, considering multiple types of stability is completed. Based on the results, the most appropriate VSC control schemes to maintain system stability are identified.
- An analysis framework is developed in combination with a hybrid modelling approach to accurately calculate the system loadability limit but with less simulation time needed compared with the conventional modelling method. The proposed framework determines which critical components must be modelled with ordinary differential equations (ODEs), with the rest of the network represented by conventional algebraic equations. The developed analysis methodology uses a considerably smaller system matrix and hence has a smaller computational burden.

1.8 Thesis Overview

This thesis consists of seven chapters overall. The six chapters that follow this introduction chapter are outlined below:

Chapter 2—*Power System Modelling and Analysis Techniques*

This chapter presents the power system modelling basis and analysis techniques employed in this thesis. The mathematical modelling of conventional power system components and VSCs, along with their control systems, is presented in the chapter initially. Following this, both linear (modal analysis) and nonlinear (bifurcation theory) analysis techniques used to assess system stability are discussed. A brief example of system stability analysis based on bifurcation theory is then included to show the effectiveness of bifurcation theory-based approaches in terms of analysing system stability limits. Finally, the test systems, alongside the software and simulation tools used throughout this thesis, are outlined.

Chapter 3—*Loadability Analysis with the Consideration of Multiple Bifurcations*

An extensive analysis of system loadability based on bifurcation theory considering voltage stability is presented in this chapter. At the start of this chapter, the necessity to update the system loadability concept is highlighted by the simultaneous consideration of multiple types of bifurcation. Following this, five loadability cases, formed of three existing and two new cases based on the updated loadability concept, are illustrated on test systems. The results demonstrate that the boundary curve based on the updated system loadability concept is a mix of various types of bifurcations, making it hard to represent using a generalised mathematical equation. To improve such composite loadability boundary, a new, multi-stage, optimisation framework is proposed to improve system stability, taking several different bifurcations into account simultaneously. The continuation method is then adopted in this chapter to verify the results from the proposed optimisation-based framework and shows that the proposed method has the capacity to optimise the overall loadability of the system in a given system parameter space.

Chapter 4—Identifying Feasibility Region Boundaries in Power Systems with Multiple VSCs

A methodology to efficiently identify and statistically analyse the system's multi-dimensional (i.e., N-dimensional) feasibility region boundary with regard to small-disturbance stability considering multiple VSCs is presented in this chapter. The need to consider system feasibility regions with N parameters varying simultaneously, particularly when multiple VSCs are connected to the system, is first discussed in the chapter. To address this, a methodology based on an interpolation technique to numerically identify the system's feasibility boundary in N domains is then presented. After the multi-dimensional feasibility region is obtained, the global sensitivity analysis technique, the partial Spearman correlation coefficient, is employed to quantify the strength of the relationships among parameters. Based on the calculated coefficient, the methodology proposed in this chapter is used to statistically identify the critical parameters affecting the system feasibility region, as well as reveal the interaction among VSCs. The method is first demonstrated on a small test system to show its effectiveness before being further applied in a large system to illustrate its generality.

Chapter 5—Identification of Critical Parameters Affecting System Loadability

This chapter presents a thorough assessment of the influential parameters and evaluation of the performance of VSC controllers over a wide range of loadability. The chapter begins with a discussion about the need to consider small-disturbance and voltage stability simultaneously during system stability analysis, particularly when a large number of VSCs are integrated into the network. The proposed methodology, which combines quasi-probabilistic simulations and a global sensitivity technique to identify influential parameters affecting the system loadability, is then laid out. The method proposed can also be used to assess VSC controller performance to establish the control system to maintain the system's largest feasible operating range. This method not only considers the variations in system parameters, such as the transmission line impedance and VSC controller parameters but also accounts for different grid strengths. The method is demonstrated in a large system with multiple VSCs.

Chapter 6—Analysing System Loadability with Multiple VSCs Using a Hybrid Model

This chapter proposes an analysis framework based on the hybrid network model to assess system loadability more efficiently. The developed hybrid network model significantly reduces the dimensions of the system matrix required to perform system loadability assessments, while still accurately identifying stability limits along the system loadability trajectory. Eigenvalue sensitivity, combined with the participation factor analysis, is adopted to identify the critical network components that require detailed modelling. This method is initially demonstrated on a small test system to show the advantages of using the hybrid model during the system loadability analysis. Following this, the proposed analysis framework is applied to large test systems with multiple VSCs. The results show that the hybrid model proposed can accurately identify various power issues using a reduced dimensional matrix as compared with other network models, which further demonstrates its effectiveness.

Chapter 7—Conclusion and Future Work

This chapter summarises and discusses the main conclusions of the research presented across the previous five chapters. Suggestions are provided for future development and improvement of the presented methodologies.

References

1. Dave Jones AS, Buck M, Graichen P (2017) The European power sector in 2017-state of affairs and review of current developments. In: Sandbag Climate Campaign & Agora Energiewende, London, UK, Rep.128/02-A-2018/EN
2. A European Green Deal, European Commission [Online]. Available: https://commission.europa.eu/strategy-and-policy/priorities-2019-2024/european-green-deal_en. Accessed 23 Jan 2023
3. Davey E, Jones C, Foster A, Ewing F (2013) UK renewable energy roadmap update 2013. Department of Energy & Climate Change, London, UK, Rep. URN 13d/259
4. Oni OE, Davidson IE, Mbangula KNI (2016) A review of LCC-HVDC and VSC-HVDC technologies and applications. In: EEEIC 2016—international conference on environment and electrical engineering
5. Lee Y, Cui S, Kim S, Sul SK (2014) Control of hybrid HVDC transmission system with LCC and FB-MMC. 2014 IEEE Energy Convers Congr Expo ECCE 2014:475–482
6. Arrillaga J, Liu YH, Watson NR (2007) Flexible power transmission: the HVDC options. Wiley
7. National Grid (2013) High voltage direct current electricity—technical information. National Grid, London, Rep.CRFS09/08/13
8. Hatziargyriou N et al (2021) Definition and classification of power system stability—revisited & extended. IEEE Trans Power Syst 36(4):3271–3281
9. Du W, Fu Q, Wang X, Wang HF (2018) Small-signal stability analysis of integrated VSC-based DC/AC power systems—a review. Int J Electr Power Energy Syst 103:545–552
10. (Prabha) Kundur P, Balu NJ, Lauby MG (1994) Power system stability and control, 1st edn. McGraw-Hill
11. Kundur P et al (2004) Definition and classification of power system stability. IEEE Trans Power Syst 19(3):1387–1401
12. Hatziargyriou N et al (2020) Stability definitions and characterization of dynamic behavior in systems with high penetration of power electronic interfaced technologies. In: IEEE Power & Energy Society, USA, Technical Report, TR-0200 (4230-46)-3
13. Padiyar KR (1999) Analysis of subsynchronous resonance in power systems, 1st edn. Springer, US, Boston
14. Jing T, Maklakov AS (2018) A review of voltage source converters for energy applications. In: Proceedings of 2018 International Ural Conference on Green Energy, Ural. 2018, pp 275–281

15. Anadon LD et al (2021) Net zero review final report: analysis exploring the key issues. HM Treasury, London
16. Skidmore C et al (2022) Mission zero—independent review of net zero. Department for Energy Security and Net Zero and Department for Business, Energy & Industrial Strategy, London
17. Slye F (2022) Future energy scenarios 2022. National Grid ESO, London
18. Ming Y, Gengyin L, Guangkai L, Haifeng L, Ming Z (2004) Modeling of VSC-HVDC and its active power control scheme. In: 2004 international conference on power system technology, POWERCON 2004, vol 2, pp 1351–1355
19. Reed G, Pape R, Takeda M (2003) "Advantages of Voltage Sourced Converter (VSC) Based Design Concepts for FACTS and HVDC-Link Applications", in 2003 IEEE Power Engineering Society General Meeting. Conf Proc 3:1816–1821
20. Clark G (2017) Clean growth strategy-leading the way to a low carbon future. Department for Business, Energy & Industrial Strategy, London
21. Interconnectors | Ofgem." [Online]. Available: https://www.ofgem.gov.uk/energy-policy-and-regulation/policy-and-regulatory-programmes/interconnectors. Accessed 26 Dec 2022
22. Edris AA et al (1997) Proposed terms and definitions for Flexible AC Transmission System (FACTS). IEEE Trans Power Deliv 12(4):1848–1853
23. Arrillaga J (1998) High voltage direct current transmission, 2nd edn. Institution of Engineering and Technology
24. Barnes M, Beddard A (2012) Voltage source converter HVDC links—the state of the art and issues going forward. Energy Procedia 24:108–122
25. Facchin C et al (2014) Special Report 60 years of HVDC. ABB Rev
26. Antonopoulos A, Angquist L, Nee H-P (2009) On dynamics and voltage control of the modular multilevel converter. In: 2009 13th European conference on power electronics and applications
27. Gnanarathna UN, Gole AM, Jayasinghe RP (2011) Efficient modeling of modular multilevel HVDC converters (MMC) on electromagnetic transient simulation programs. IEEE Trans Power Deliv 26(1):316–324
28. Tu Q, Xu Z, Xu L (2011) Reduced Switching-frequency modulation and circulating current suppression for modular multilevel converters. IEEE Trans Power Deliv 26(3):2009–2017
29. Gleick J 1987 Chaos: making a new science, 1st edn. Viking Books
30. Strogatz SH (2000) Nonlinear dynamics and chaos: applications to physics, biology, chemistry, and engineering, 1st edn. CRC Press
31. Phillips CL, Parr JM, Riskin EA (2008) Signals, systems, and transforms, 5th edn. Pearson Education
32. Pichery C (2014) Sensitivity analysis. In: Encyclopedia of toxicology, 3rd edn. Academic Press, pp 236–237
33. Kasturi R, Dorraju P (1969) Sensitivity analysis of power systems. IEEE Trans Power Appar Syst PAS-88 10:1–23
34. Konishi H, Takahashi C, Kishibe H, Sato H (2002) A consideration of stable operating power limits in VSC-HVDC systems. IEE Conf Publ 485:102–106
35. Durrant M, Werner H, Abbott K (2003) Model of a VSC HVDC terminal attached to a weak AC system. IEEE Conf Control Appl Proc 1:178–182
36. Jovcic D, Lamont LA, Xu L (2003) VSC transmission model for analytical studies. 2003 IEEE Power Eng Soc Gen Meet Conf Proc 3:1737–1742
37. Rezaee S, Radwan A, Moallem M, Wang J (2021) Dual active compensation for voltage source rectifiers under very weak grid conditions. IEEE Access 9:160446–160460
38. Wang D, Hou Y, Hu J (2019) Comparative analysis of stability limitations in weak grid-connected synchronous generator, VSC, and DFIG systems considering the power flow control dynamics. IET Renew Power Gener 13(1):94–102
39. Wu G, Wang S, Zhang X, Zhao B, Li Y, Wang T (2020) Analysis on the low-frequency bifurcation phenomena of weak-grid-tied VSCs. In: IEEE international symposium on circuits and systems

40. Liserre M, Teodorescu R, Blaabjerg F (2006) Stability of photovoltaic and wind turbine grid-connected inverters for a large set of grid impedance values. IEEE Trans Power Electron 21(1):263–271
41. Guo J et al (2021) Wideband dq-frame impedance modeling of load-side virtual synchronous machine and its stability analysis in comparison with conventional PWM rectifier in weak grid. IEEE J Emerg Sel Top Power Electron 9(2):2440–2451
42. Zhou Y, Nguyen DD, Kjaer PC, Saylors S (2013) Connecting wind power plant with weak grid—challenges and solutions. In: IEEE power and energy society general meeting
43. Rezaee S, Radwan A, Moallem M, Wang J (2020) Voltage source converters connected to very weak grids: accurate dynamic modeling, small-signal analysis, and stability improvement. IEEE Access 8:201120–201133
44. Zhou JZ, Ding H, Fan S, Zhang Y, Gole AM (2014) Impact of short-circuit ratio and phase-locked-loop parameters on the small-signal behavior of a VSC-HVDC converter. IEEE Trans Power Deliv 29(5):2287–2296
45. Dong D, Li J, Boroyevich D, Mattavelli P, Cvetkovic I, Xue Y (2012) Frequency behavior and its stability of grid-interface converter in distributed generation systems. In: 2012 twenty-seventh annual IEEE Applied Power Electronics Conference and Exposition (APEC)
46. Dong D, Wen B, Boroyevich D, Mattavelli P, Xue Y (2015) Analysis of phase-locked loop low-frequency stability in three-phase grid-connected power converters considering impedance interactions. IEEE Trans Ind Electron 62(1):310–321
47. Wen B, Boroyevich D, Burgos R, Mattavelli P, Shen Z (2016) Analysis of D-Q small-signal impedance of grid-tied inverters. IEEE Trans Power Electron 31(1):675–687
48. Wen B, Boroyevich D, Mattavelli P, Shen Z, Burgos R (2013) Influence of phase-locked loop on input admittance of three-phase voltage-source converters. In: 2013 Twenty-Eighth Annual IEEE Applied Power Electronics Conference and Exposition (APEC)
49. Zhang Y, Tian M, Zhang H, Song J, Zhang W (2022) Admittance modeling and stability enhancement of grid-connected inverter considering frequency coupling in weak grids. Electr Power Syst Res 209:108034
50. Yazdani A, Iravani R (2010) Voltage-sourced converters in power systems: modeling, control, and applications, 1st edn. Wiley-IEEE Press
51. Alawasa KM, Mohamed YARI, Xu W (2014) Active mitigation of subsynchronous interactions between PWM voltage-source converters and power networks. IEEE Trans Power Electron 29(1):121–134
52. Zhou P, Yuan X, Hu J, Huang Y (2014) Stability of DC-link voltage as affected by phase locked loop in VSC when attached to weak grid. In: IEEE power and energy society general meeting
53. Huang Y, Yuan X, Hu J, Zhou P (2015) Modeling of VSC connected to weak grid for stability analysis of DC-link voltage control. IEEE J Emerg Sel Top Power Electron 3(4):1193–1204
54. Huang Y, Wang D (2018) Effect of control-loops interactions on power stability limits of VSC integrated to AC system. IEEE Trans Power Deliv 33(1):301–310
55. Wang D, Liang L, Shi L, Hu J, Hou Y (2019) Analysis of modal resonance between PLL and DC-link voltage control in weak-grid tied VSCs. IEEE Trans Power Syst 34(2):1127–1138
56. Morris JF, Ahmed KH, Egea-Alvarez A (2021) Analysis of controller bandwidth interactions for vector-controlled VSC connected to very weak ac grids. IEEE J Emerg Sel Top Power Electron 9(6):7343–7354
57. Midtsund T, Suul JA, Undeland T (2010) Evaluation of current controller performance and stability for voltage source converters connected to a weak grid. In: 2nd international symposium on power electronics for distributed generation systems, PEDG 2010, pp 382–388
58. Zhao M, Yuan X, Hu J, Yan Y (2016) Voltage dynamics of current control time-scale in a VSC-connected weak grid. IEEE Trans Power Syst 31(4):2925–2937
59. Liu H, Xie X (2019) Comparative studies on the impedance models of VSC-based renewable generators for SSI stability analysis. IEEE Trans Energy Convers 34(3):1442–1453
60. Xue T, Lyu J, Li Y, Wang H, Cai X (2020) A comprehensive study on impedance models of grid-tied voltage-source converters. In: IECON proceedings (industrial electronics conference)

61. Zhou W, Mohammed N, Bahrani B (2022) Comprehensive modeling, analysis, and comparison of state-space and admittance models of PLL-based grid-following inverters considering different outer control modes. IEEE Access 10:30109–30146
62. Zong H et al (2018) Analysis of bifurcation behaviors in MMC connected to a weak grid. In: Proceedings: IECON 2018—44th annual conference of the IEEE industrial electronics society, pp 1687–1692
63. Huang M, Tse CK, Wong SC, Wan C, Ruan X (2013) Low-frequency hopf bifurcation and its effects on stability margin in three-phase PFC power supplies connected to non-ideal power grid. IEEE Trans Circuits Syst I Regul Pap 60(12):3328–3340
64. Huang M, Wong S-C, Tse CK, Ruan X (2013) Catastrophic bifurcation in three-phase voltage-source converters. IEEE Trans Circuits Syst I Regul Pap 60(4)
65. Arani MFM, Mohamed YARI (2017) Analysis and performance enhancement of vector-controlled VSC in HVDC links connected to very weak grids. IEEE Trans Power Syst 32(1):684–693
66. Alawasa KM, Mohamed YARI, Xu W (2013) Modeling, analysis, and suppression of the impact of full-scale wind-power converters on subsynchronous damping. IEEE Syst J 7(4):700–712
67. Sun Z, Zhao J, Ma S (2018) Steady-state voltage stability analysis of AC-DC hybrid power system with VSC-HVDC. In: 2018 2nd IEEE conference on energy internet and energy system integration (EI2)
68. Zhao L, Zheng H (2019) Voltage stability analysis for AC/DC system with VSC-HVDC. In: 2019 2nd international conference on electronics technology, pp 195–199
69. Du W, Chen X, Wang HF (2017) Impact of dynamic interactions introduced by the DFIGs on power system electromechanical oscillation modes. IEEE Trans Power Syst 32(6):4954–4967
70. Du W, Chen X, Wang HF (2017) Strong dynamic interactions of grid-connected DFIGs with power systems caused by modal coupling. IEEE Trans Power Syst 32(6):4386–4397
71. Tian X, Tang H, Li Y, Chi Y, Su Y (2017) Dynamic stability of weak grid connection of large-scale DFIG based on wind turbines. J Eng 2017(13):1092–1097
72. Du W, Chen X, Wang HF (2017) Power system electromechanical oscillation modes as affected by dynamic interactions from grid-connected PMSGs for wind power generation. IEEE Trans Sustain Energy 8(3):1301–1312
73. Zhu L et al (2017) Analysis of interaction between HVDC and offshore wind power in receiving end grid. In: 2017 IEEE conference on energy internet and energy system integration (EI2), pp 1–6
74. Yuan H, Yuan X, Hu J (2017) Modeling of grid-connected VSCs for power system small-signal stability analysis in DC-link voltage control timescale. IEEE Trans Power Syst 32(5):3981–3991
75. Harnefors L, Bongiorno M, Lundberg S (2007) Input-admittance calculation and shaping for controlled voltage-source converters. IEEE Trans Ind Electron 54(6):3323–3334
76. Pogaku N, Prodanović M, Green TC (2007) Modeling, analysis and testing of autonomous operation of an inverter-based microgrid. IEEE Trans Power Electron 22(2):613–625
77. Shen L, Barnes M, Milanović JV, Bell KRW, Belivanis M (2014) Potential interactions between VSC HVDC and STATCOM. In: Proceedings of 2014 power systems computation conference, PSCC
78. Grdenic G, Delimar M, Beerten J (2019) Comparative analysis on small-signal stability of multi-infeed VSC HVDC system with different reactive power control strategies. IEEE Access 7:151724–151732
79. Alawasa KM (2014) Modeling, analysis and mitigation of sub-synchronous interactions between full- and partial-scale voltage-source converters and power networks. Department of Electrical Engineering and Energy, University of Alberta, Edmonton, Alberta
80. Wan C, Huang M, Tse CK, Ruan X (2015) Effects of interaction of power converters coupled via power grid: a design-oriented study. IEEE Trans Power Electron 30(7):3589–3600
81. Chen H, Yu W, Liu Z, Yan Q, Adamu Tasiu I, Han Z (2020) Low-frequency instability induced by Hopf bifurcation in a single-phase converter connected to non-ideal power grid. IEEE Access 8:62871–62882

82. Shah R, Barnes M, Preece R (2017) Characterisation of interaction in an offshore AC grid with large WPPs and VSC-HVDCs. IET Renew Power Gener 11(6):723–731
83. Bayo-Salas A, Van Hertem D, Rimez J, Beerten J (2016) Analysis of control interactions in multi-infeed VSC HVDC connections. IET Gener Transm Distrib 10(6):1336–1344
84. Huang Y, Wang D, Shang L, Zhu G, Tang H, Li Y (2017) Modeling and stability analysis of DC-link voltage control in multi-VSCs with integrated to weak grid. IEEE Trans Energy Convers 32(3):1127–1138
85. Hu J, Huang Y, Wang D, Yuan H, Yuan X (2015) Modeling of grid-connected DFIG-based wind turbines for dc-link voltage stability analysis. IEEE Trans. Sustain. Energy 6(4):1325–1336
86. Wang X, Blaabjerg F, Wu W (2014) Modeling and analysis of harmonic stability in an AC power-electronics- based power system. IEEE Trans Power Electron 29(12):6421–6432
87. Givaki K, Xu L (2015) Stability analysis of large wind farms connected to weak AC networks incorporating PLL dynamics. In: IET conference on renewable power generation, vol 2015, no CP679
88. Yang D, Wang X, Blaabjerg F (2018) Sideband harmonic instability of paralleled inverters with asynchronous carriers. IEEE Trans Power Electron 33(6):4571–4577
89. Roose T, Lekic A, Alam MM, Beerten J (2021) Stability analysis of high-frequency interactions between a converter and HVDC grid resonances. IEEE Trans Power Deliv 36(6):3414–3425
90. Choi B (1995) A method of defining the load impedance specification for a stable distributed power system. IEEE Trans Power Electron 10(3):280–285
91. Tavakoli SD, Prieto-Araujo E, Sánchez-Sánchez E, Gomis-Bellmunt O (2020) Interaction assessment and stability analysis of the MMC-based VSC-HVDC link. Energies 13(8):2075
92. Yu G, Xu J, Yan Z, Chen W, Zhao D, Zhang Z (2021) Stability optimization of low-voltage DC flexible interconnected power station in AC distribution network based on impedance analysis. In: 2021 international conference on power system technology, Carbon Neutrality New Type Power System. POWERCON 2021, pp 1659–1666
93. Song Y, Breitholtz C (2016) Nyquist stability analysis of an AC-grid connected VSC-HVDC system using a distributed parameter DC cable model. IEEE Trans Power Deliv 31(2):898–907
94. Stamatiou G, Bongiorno M (2016) Stability analysis of two-terminal VSC-HVDC systems using the net-damping criterion. IEEE Trans Power Deliv 31(4):1748–1756
95. Trinh NT, Zeller M, Wuerflinger K, Erlich I (2016) Generic model of MMC-VSC-HVDC for interaction study with AC power system. IEEE Trans Power Syst 31(1):27–34
96. Gontijo GF, Bakhshizadeh MK, Kocewiak LH, Teodorescu R (2022) State space modeling of an offshore wind power plant with an MMC-HVDC connection for an eigenvalue-based stability analysis. IEEE Access 10:82844–82869
97. Mochamad RF, Preece R (2020) Assessing the impact of VSC-HVDC on the interdependence of power system dynamic performance in uncertain mixed AC/DC systems. IEEE Trans Power Syst 35(1):63–74
98. Pipelzadeh Y, Chaudhuri B, Green TC (2011) Stability improvement through HVDC upgrade in the Australian equivalent system. In: 45th International Universities Power Engineering Conference UPEC2010
99. Pipelzadeh Y, Chaudhuri B, Green TC (2010) Wide-area power oscillation damping control through HVDC: a case study on Australian equivalent system. In: IEEE PES general meeting
100. Chaudhuri N, Chaudhuri B, Majumder R, Yazdani A (2014) Multi-terminal direct-current grids: modeling, analysis, and control, 1st edn. Wiley-IEEE Press
101. D'Arco S, Beerten J, Suul JA (2017) Classification and analysis of impact on small-signal dynamics and stability from expansion of VSC-HVDC systems to multi-terminal HVDC grids. In: 13th IET international conference on AC and DC power transmission (ACDC 2017)
102. Beerten J, D'Arco S, Suul JA (2016) Identification and small-signal analysis of interaction modes in VSC MTDC systems. IEEE Trans Power Deliv 31(2):888–897
103. Du W, Fu Q, Wang H (2018) Small-signal stability of an AC/MTDC power system as affected by open-loop modal coupling between the VSCs. IEEE Trans Power Syst 33(3):3143–3152

104. Du W, Fu Q, Wang H (2017) Strong dynamic interactions between multi-terminal DC network and AC power systems caused by open-loop modal coupling. IET Gener Transm Distrib 11(9):2362–2374
105. Dewangan L, Bahirat HJ (2020) Controller interaction and stability margins in mixed SCR MMC-based HVDC Grid. IEEE Trans Power Syst 35(4):2835–2846
106. Chaudhuri NR, Majumder R, Chaudhuri B, Pan J (2011) Stability analysis of VSC MTDC grids connected to multimachine AC systems. IEEE Trans Power Deliv 26(4):2774–2784
107. Suriyaarachchi DHR, Annakkage UD, Karawita C, Jacobson DA (2013) A procedure to study sub-synchronous interactions in wind integrated power systems. IEEE Trans Power Syst 28(1):377–384
108. Suriyaarachchi DHR, Annakkage UD, Karawita C, Kell D, Mendis R, Chopra R (2012) Application of an SVC to damp sub-synchronous interaction between wind farms and series compensated transmission lines. In: 2012 IEEE power and energy society general meeting
109. Pasrija AK, Bahar LY (1986) Static bifurcations in electric power networks: loss of steady-state stability and voltage collapse. IEEE Trans Circuits Syst 33(10):981–991
110. Chiang HD, Wang CS, Flueck AJ (1997) Look-ahead voltage and load margin contingency selection functions for large-scale power systems. IEEE Trans Power Syst 12(1):173–180
111. Van Cutsem T (1991) A method to compute reactive power margins with respect to voltage collapse. IEEE Trans Power Syst 6(1):145–156
112. Dobson I, Lu L (1992) Voltage collapse precipitated by the immediate change in stability when generator reactive power limits are encountered. IEEE Trans Circuits Syst I Fundam Theory Appl 39(9):762–766
113. Cañizares CA, Hranilovic S (1994) Transcritical and Hopf bifurcations in AC/DC systems. In: Proceedings of bulk power system voltage phenomena—III seminar on voltage stability, security and control, pp 105–114
114. Alvarado FL, Jung TH (1989) Direct detection of voltage collapse conditions. In: Bulk power system voltage phenomena: voltage stability and security
115. Canizares CA, Alvarado FL, DeMarco CL, Dobson I, Long WF (1992) Point of collapse methods applied to AC/DC power systems. IEEE Trans Power Syst 7(2):673–683
116. Cañizares CA, Alvarado FL (1993) Point of collapse and continuation methods for large AC/DC systems. IEEE Trans Power Syst 8(1):1–8
117. Casteroba Bento ME, Andrade Ramos R (2018) A method for dynamic security assessment of power systems with simultaneous consideration of Hopf and Saddle-node bifurcations. In: IEEE power and energy society general meeting
118. Ajjarapu V (2007) Computational techniques for voltage stability assessment and control, 1st edn. Springer, Berlin
119. Ajjarapu V, Christy C (1992) The continuation power flow: a tool for steady state voltage stability analysis. IEEE Trans Power Syst 7(1):416–423
120. Bedriñana MF, Castro CA (2008) Maximum loading point computation based on load flow with step size optimization. In: IEEE power and energy society 2008 general meeting: conversion and delivery of electrical energy in the 21st century
121. Vesti S, Suntio T, Oliver JA, Prieto R, Cobos JA (2013) Impedance-based stability and transient-performance assessment applying maximum peak criteria. IEEE Trans Power Electron 28(5):2099–2104
122. Irisarri GD, Wang X, Tong J, Mokhtari S (1997) Maximum loadability of power systems using interior point non-linear optimization method. IEEE Trans Power Syst 12(1):162–172
123. Parker CJ, Morrison IF, Sutanto D (1996) Application of an optimisation method for determining the reactive margin from voltage collapse in reactive power planning. IEEE Trans Power Syst 11(3):1473–1481
124. Li P, Qi J, Wang J, Wei H, Bai X, Qiu F (2017) An SQP method combined with gradient sampling for small-signal stability constrained OPF. IEEE Trans Power Syst 32(3):2372–2381
125. Makarov YV, Dong ZY, Hill DJ (2002) A general method for small signal stability analysis [of power systems]. In: 20th international conference on power industry computer applications, pp 280–286

126. Hill DJ, Dong ZY (2000) Nonlinear computation and control for small disturbance stability. IEEE Power Eng Soc Transm Distrib Conf 2:836–841
127. Avalos RJ, Cañizares CA, Milano F, Conejo AJ (2009) Equivalency of continuation and optimization methods to determine saddle-node and limit-induced bifurcations in power systems. IEEE Trans Circuits Syst I Regul Pap 56(1):210–223
128. Gu X, Cañizares CA (2006) Fast prediction of loadability margins by constructing a small-signal stability boundary based on neural networks. In: 2006 international conference on power system technology
129. Gutierrez-Martinez VJ, Cañizares CA, Fuerte-Esquivel CR, Pizano-Martinez A, Gu X (2011) Neural-network security-boundary constrained optimal power flow. IEEE Trans Power Syst 26(1):63–72
130. Sarkar MNI, Meegahapola LG, Datta M (2018) Reactive power management in renewable rich power grids: a review of grid-codes, renewable generators, support devices, control strategies and optimization Algorithms. IEEE Access 6:41458–41489
131. Sun Y et al (2023) Optimization methods for optimal power flow problems in distribution networks: a brief review. In: 2023 8th Asia Conference on Power and Electrical Engineering (ACPEE), pp 1400–1406
132. Thirunavukkarasu M, Sawle Y, Lala H (2023) A comprehensive review on optimization of hybrid renewable energy systems using various optimization techniques. Renew Sustain Energy Rev 176:113192
133. Behera S, Sahoo S, Pati BB (2015) A review on optimization algorithms and application to wind energy integration to grid. Renew Sustain Energy Rev 48:214–227
134. Ayalew F, Hussen S, Pasam DG (2018) Optimization techniques in power system: review, vol 3, pp 8–16
135. Cañizares CA (1998) Applications of optimization to voltage collapse analysis. In: IEEE PES summer meeting
136. Yorino N, Li HQ, Sasaki H (2005) A predictor/corrector scheme for obtaining Q-limit points for power flow studies. IEEE Trans Power Syst 20(1):130–137
137. Hiskens IA, Chakrabarti BB (1996) Direct calculation of reactive power limit points. Int J Electr Power Energy Syst 18(2):121–129
138. Jalali A, Aldeen M (2018) Efficient method to identify saddle-node and limit-induced bifurcation points of power system. IET Gener Transm Distrib 12(17):4034–4044
139. Rosehart W, Roman C, Schellenberg A (2008) Optimal power flow with complementarity constraints. IEEE Trans Power Syst 20(2):417–417
140. Rosehart W, Canizares C, Quintana V (1999) Optimal power flow incorporating voltage collapse constraints. IEEE Power Eng Soc Summer Meeting 1999:820–825
141. Vournas CD, Karystianos M, Maratos NG (2000) Bifurcation points and loadability limits as solutions of constrained optimization problems. In: IEEE power engineering society transmission and distribution conference, pp 1883–1888
142. Karystianos ME, Maratos NG, Vournas CD (2007) Maximizing power-system loadability in the presence of multiple binding complementarity constraints. IEEE Trans Circuits Syst I Regul Pap 54(8):1775–1787
143. Van Cutsem T, Vournas CD (2007) Voltage stability of electric power system, 1st edn. Springer, Berlin
144. Neves LS, Costa Alberto LF (2020) On the computation of the locally closest bifurcation point considering loading uncertainties and reactive power limits. IEEE Trans. Power Syst 35(5):3885–3894
145. Dobson I, Lu L (1993) New methods for computing a closest saddle node bifurcation and worst case load power margin for voltage collapse. IEEE Trans Power Syst 8(3):905–913
146. Canizares CA (1998) Calculating optimal system parameters to maximize the distance to saddle-node bifurcations. IEEE Trans Circuits Syst I Fundam Theory Appl 45(3):225–237
147. Mittelstaedt M, Bauer S, Schnettler A (2017) An iterative method for determining the most probable bifurcation in large scale power systems. IEEE Trans Power Syst 32(4):2966–2973

148. Ajjarapu V, Lee B (1992) Bifurcation theory and its application to nonlinear dynamical phenomena in an electrical power system. IEEE Trans Power Syst 7(1):424–431
149. Vournas CD (1996) The effect of Automatic Voltage Regulation on the bifurcation evolution in power systems. IEEE Trans Power Syst 11(4):1683–1688
150. Zecevic AI, Miljkovic DM (2002) The effects of generation redispatch on Hopf bifurcations in electric power systems. IEEE Trans Circuits Syst I Fundam Theory Appl 49(8):1180–1186
151. Cañizares CA, Mithulananthan N, Milano F (2004) Linear performance indices to predict oscillatory stability problems in power systems. IEEE Trans Power Syst 19(2):1104–1114
152. Ghasemi H, Cañizares CA, Moshref A (2006) Oscillatory stability limit prediction using stochastic subspace identification. IEEE Trans Power Syst 21(2):736–745
153. Dobson I (1993) Computing a closest bifurcation instability in multidimensional parameter space. J Nonlinear Sci 3(1):307–327
154. Joshi SK, Srivastava SC (1996) Estimation of closest Hopf bifurcation in electric power systems. In: 12th power system computation conference, pp 195–200
155. Phadke AR, Fozdar M, Niazi KR, Mithulananthan N, Bansal RC (2011) New technique for computation of closest Hopf bifurcation point using real-coded genetic algorithm. IET Gener Transm Distrib 5(1):11–18
156. Gupta CP, Varma RK, Srivastava SC (1998) Method to determine closest Hopf bifurcation in power systems considering exciter and load dynamics. In: Proceedings 1995 international conference on energy management and power delivery, EMPD, vol 1, pp 293–297
157. Bevin KC, Krishan R, Verma A (2020) Fast identification of ancillary service locations for improving hopf bifurcation margin under uncertainties. In: 9th IEEE power India international conference
158. Revel G, Leon AE, Alonso DM, Moiola JL (2010) Bifurcation analysis on a multimachine power system model. IEEE Trans Circuits Syst I Regul Pap 57(4):937–949
159. Zadehkhost PS, Howell F, Lin X, Wang L (2017) Analyzing subsynchronous control interactions in large-scale power systems in frequency domain. In: 2017 IEEE Power & Energy Society General Meeting
160. Lerm A, Cañizares CA, Lemos FAB, e Silva AS (1998) Multi-parameter bifurcation analysis of power systems. In: North American Power Symposium
161. Lerm AAP, Cañizares CA, e. Silva AS (2003) Multiparameter bifurcation analysis of the South Brazilian power system. IEEE Trans Power Syst 18(2):737–746
162. Zhou Y, Ajjarapu V (2005) A fast algorithm for identification and tracing of voltage and oscillatory stability margin boundaries. Proc IEEE 93(5):934–946
163. Gomes S, Martins N, Portela C (2003) Computing small-signal stability boundaries for large-scale power systems. In: 2003 IEEE Power Engineering Society General Meeting
164. Gomes S, Martins N, Parreiras TJMA (2017) Computing the closest small-signal security boundary in the control parameter space for large scale power systems. Electr Power Syst Res 149:10–18
165. Shuai Z, Hu Y, Peng Y, Tu C, Shen ZJ (2017) Dynamic stability analysis of synchronverter-dominated microgrid based on bifurcation theory. IEEE Trans Ind Electron 64(9):7467–7477
166. Nam HK (2000) A new eigen-sensitivity theory of augmented matrix and its applications to power system stability analysis. IEEE Trans Power Syst 15(1):363–369
167. Li C, Chiang HD, Du Z (2017) Network-preserving sensitivity-based generation rescheduling for suppressing power system oscillations. IEEE Trans Power Syst 32(5):3824–3832
168. Li C, Chiang HD, Du Z (2018) Investigation of an effective strategy for computing small-signal security margins. IEEE Trans Power Syst 33(5):5437–5445
169. Wen X, Ajjarapu V (2006) Application of a novel eigenvalue trajectory tracing method to identify both oscillatory stability margin and damping margin. IEEE Trans Power Syst 21(2):817–824
170. Li C, Du Z (2013) A novel method for computing small-signal stability boundaries of large-scale power systems. IEEE Trans Power Syst 28(2):877–883
171. Greene S, Dobson I, Alvarado FL (1997) Sensitivity of the loading margin to voltage collapse with respect to arbitrary parameters. IEEE Trans Power Syst 12(1):262–272

172. Dobson I, Lu L (1992) Computing an optimum direction in control space to avoid saddle node bifurcation and voltage collapse in electric power systems. IEEE Trans Automat Contr 37(10):1616–1620
173. Dobson I, Alvarado F, DeMarco CL (2005) Sensitivity of Hopf bifurcations to power system parameters. In: 31st IEEE conference on decision and control, pp 2928–2933
174. Preece R, Milanović JV, Almutairi AM, Marjanovic O (2013) Probabilistic evaluation of damping controller in networks with multiple VSC-HVDC lines. IEEE Trans Power Syst 28(1):367–376
175. Hasan KN, Preece R, Milanovic JV (2017) Priority ranking of critical uncertainties affecting small-disturbance stability using sensitivity analysis techniques. IEEE Trans Power Syst 32(4):2629–2639
176. Hasan KN, Saha TK, Eghbal M (2014) Investigating the priority of market participants for low emission generation entry into the Australian grid. Energy 71:445–455
177. Preece R, Milanovic JV (2016) Assessing the applicability of uncertainty importance measures for power system studies. IEEE Trans Power Syst 31(3):2076–2084
178. Guddanti B, Orrego JR, Roychowdhury R, Illindala MS (2022) Sensitivity analysis based identification of key parameters in the dynamic model of a utility-scale solar PV plant. IEEE Trans Power Syst 37(2):1340–1350
179. Qi B, Hasan KN, Milanovic JV (2019) Identification of critical parameters affecting voltage and angular stability considering load-renewable generation correlations. IEEE Trans Power Syst 34(4):2859–2869
180. Iooss B, Lemaître P (2015) A review on global sensitivity analysis methods. In: Uncertainty management in simulation-optimization of complex systems: algorithms and applications. Springer New York LLC, pp 101–122, vol 59

Chapter 2
Power System Modelling and Analysis Techniques

This chapter details the fundamental techniques used to perform the stability analysis of power systems with multiple connected VSCs. Further analysis and in-depth study in the following chapters will be based on the fundamental modelling and analysis approaches introduced in this chapter. This chapter first discusses mathematical models of the key components in a conventional power system, including synchronous generators, excitation systems, branches, system loads, and the electrical network. The second section of this chapter addresses the techniques used to represent VSCs. An overview of various established VSC models from the CIGRE and the average-value converter model adopted in this thesis are briefly introduced. Subsequently, key elements that influence the dynamic performance of VSCs are explored, including the phase-locked loops, active and reactive power control, and AC voltage control. Following the introduction of the modelling of each power system component, the completed overall power system model with all components connected will be presented.

Alongside the included modelling methods, the power system analysis tools, which are based on either system linearisation or bifurcation theory, are described, and an example of power system stability analysis using bifurcation theory will be presented. Finally, a description of the various test networks used in this thesis will be introduced.

Throughout this thesis, all modelling and coding have been completed in either MATLAB® (2019a), for analytical analysis, or Simulink® (2020a) and DIgSILENT (2022) for EMT time-domain simulation.

Y. Chen, *Stability Assessment of Power Systems with Multiple Voltage Source Converters*, Springer Theses, https://doi.org/10.1007/978-3-031-63095-8_2

2.1 Modelling Power System Components

This section will introduce mathematical models for all the major power system components in a conventional power system. The models represented in this section will then be adopted to perform power system stability analysis in the following chapters. It should be noted that the studies are performed on power transmission systems. Hence, the power system components are modelled using an orthogonal phase representation and assuming that all three phases are balanced.

2.1.1 *The Synchronous Generator*

The synchronous generator (SG) is one of the critical energy sources on the AC side of the network. Figure 2.1 illustrates the schematic of an SG consisting of two components: a rotor and a stator. The excitation voltage is applied to the rotor winding, which is also known as the field winding, to generate an excitation current and induce a magnetic field. The stator windings are spread evenly in the space, with 120° between each phase, as also shown in Fig. 2.1. With the rotor rotating at an angular speed ω_r, the magnetic field generated in the field winding will continuously cut the stator windings in three phases and hence, under balanced conditions, induce an AC voltage in each phase with a phase difference of 120°. Compared with the conventional three-phase (*abc*) frame, the SG is typically modelled in the *dq* domain to clearly illustrate the SG's dynamic behaviour and simplify the calculation. This is achieved using an $abc - dq$ transformation based on the rotor angle. The rotor angle, δ, in Fig. 2.1 represents the angle of the d-axis of the rotor, with the phase angle of the a-axis providing the reference value. In general, the $abc - dq$ transformation can be viewed as an approach to refer variables in the stator winding to the rotor side. With the SG modelled in the *dq* domain, the time-variant stator quantities are converted to constant values under balanced steady-state conditions.

Numerous well-documented SG models considering various modelling details can be found in previous literature. The sixth-order SG model [1, 2] will be first illustrated. This is followed by a simplified fourth-order SG model [2] which neglects the dynamics in the sub-transient circuit and which is adopted in this thesis.

2.1.1.1 The Detailed Sixth-Order Model

The first-order differential equations for the sixth-order SG model are provided in (2.1)–(2.6).

$$\frac{\mathrm{d}\delta}{\mathrm{d}t} = (\omega - \omega_{\mathrm{ref}})\omega_0 \tag{2.1}$$

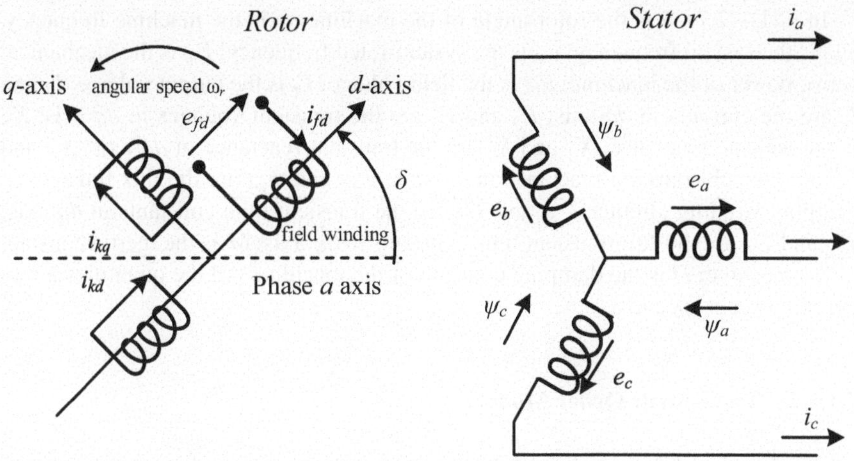

Fig. 2.1 A synchronous machine circuit

$$M\frac{\mathrm{d}\omega}{\mathrm{d}t} = P_m - D(\omega - \omega_{\mathrm{ref}}) - \frac{(X_d'' - X_{ls})}{(X_d' - X_{ls})}E_q' I_q - \frac{(X_d' - X_d'')}{(X_d' - X_{ls})}\psi_{1d} I_q$$
$$- \frac{\left(X_q'' - X_{ls}\right)}{\left(X_q' - X_{ls}\right)}E_d' I_d - \frac{\left(X_q' - X_q''\right)}{\left(X_q' - X_{ls}\right)}\psi_{2d} I_q - \left(X_q'' - X_d''\right)I_d I_q \tag{2.2}$$

$$T_{d0}'\frac{\mathrm{d}E_q'}{\mathrm{d}t} = E_{fd} - E_q' - (X_d - X_d')\left[I_d - \frac{(X_d' - X_d'')}{(X_d' - X_{ls})^2}(\psi_{1d} + (X_d' - X_{ls})I_d - E_q')\right] \tag{2.3}$$

$$T_{q0}'\frac{\mathrm{d}E_d'}{\mathrm{d}t} = -E_d' + (X_q - X_q')\left[I_q - \frac{\left(X_q' - X_q''\right)}{\left(X_q' - X_{ls}\right)^2}(\psi_{2d} + \left(X_q' - X_{ls}\right)I_q + E_d')\right] \tag{2.4}$$

$$T_{d0}''\frac{\mathrm{d}\psi_{1d}}{\mathrm{d}t} = -\psi_{1d} + E_q' - (X_d' - X_{ls})I_d \tag{2.5}$$

$$T_{q0}''\frac{\mathrm{d}\psi_{2d}}{\mathrm{d}t} = -\psi_{2d} - E_d' - (X_q' - X_{ls})I_q \tag{2.6}$$

The algebraic equation defining the interface voltage between the SG and the network is given by (2.7).

$$E_t = \sqrt{E_d^2 + E_q^2} \tag{2.7}$$

In (2.1)–(2.7), δ is the rotor angle of the machine; ω is the machine frequency, ω_{ref} is the system frequency; ω_0 is the system-rated frequency; P_m is the mechanical active power of the machine; E_{fd} is the field voltage; E_t is the output voltage; I_d and I_q are the currents in dq-axis; E_d' and E_q' are the transient voltages in dq-axis; X_{ls} is the leakage reactance; X_d' and X_q' are the transient reactance in dq-axis; X_d'' and X_q'' are the sub-transient reactance in dq-axis; ψ_{1d} and ψ_{2d} are the flux linkages in damping winding circuits; T_{d0}' and T_{q0}' are the transient time constants in dq-axis; T_{d0}'' and T_{q0}'' are the sub-transient time constants in dq-axis; M is the inertia constant of the machine; D is the damping constant of the machine; All the quantities are in per unit except ω_0 and δ.

2.1.1.2 The Fourth-Order Model

As discussed in Chap. 1, local bifurcation theory can be used to analyse power system small-disturbance stability but is unable to be employed to investigate a power system's transient stability. Accordingly, the fourth-order SG model neglects the dynamics in the SG damper winding circuit (i.e. ψ_{1d} and ψ_{2d}), which has a significant impact on the SG's transient stability but a limited effect on the small-disturbance stability of the system [2]. Additionally, considering the frequency of interest within this thesis, which extends slightly above the fundamental frequency, the fourth-order SG model is chosen. The fourth-order model is a typical SG model used in research where local bifurcation theory has been used to analyse power system small-disturbance stability [3–5]. Assuming that the time constants of the damper winding (T_{d0}'' and T_{q0}'') are sufficiently small (i.e. equal to zero in this case), the generator state equations are given by (2.8)–(2.11).

$$\frac{\mathrm{d}\delta}{\mathrm{d}t} = (\omega - \omega_{\mathrm{ref}})\omega_0 \tag{2.8}$$

$$M\frac{\mathrm{d}\omega}{\mathrm{d}t} = P_m - D(\omega - \omega_{\mathrm{ref}}) - \left(E_q' - X_d' I_d\right)I_q - \left(E_d' + X_q' I_q\right)I_d \tag{2.9}$$

$$T_{d0}'\frac{\mathrm{d}E_q'}{\mathrm{d}t} = E_{fd} - E_q' - (X_d - X_d')I_d \tag{2.10}$$

$$T_{q0}'\frac{\mathrm{d}E_d'}{\mathrm{d}t} = -E_d' + (X_q - X_q')I_q \tag{2.11}$$

It should be noted that one of the generators will be selected as the reference machine to provide the system angle reference (its angle is deemed to be the system angle reference). Consequently, (2.8)–(2.9) are neglected in the reference SG modelling. The algebraic equations describing the generator stator voltage in the dq-axis are defined as (2.12)–(2.13).

$$0 = -E'_q + V_d^{sg} + X'_d I_d \tag{2.12}$$

$$0 = -E'_d + V_q^{sg} - X'_q I_q \tag{2.13}$$

In (2.12)–(2.13), V_d^{sg} and V_q^{sg} are the SG's terminal voltages in the d and q axes.

2.1.2 The Excitation System

Excitation systems regulate generator terminal voltage and improve system stability, particularly first-swing stability. The excitation system is interfaced with the SG with the generator field voltage. By controlling the field voltage (and therefore the field current in the field winding), the SG's terminal voltage is adjusted. This is provided by an automatic voltage regulator (AVR), which manipulates the field voltage (and therefore the field current in the rotor's field winding) to its set reference point. Figure 2.2 shows the block diagram clarifying the functional relationship between the SG and the excitation system.

Various excitation systems have been developed and used in practice. A thorough and comprehensive explanation of excitation systems can be found in [6, 7]. Among these, IEEE Type DC1A is used in this thesis, as presented in Fig. 2.3. This model is adopted to represent the excitation system, which uses the controlled field DC current as a means of regulating the generator terminal voltage. Figure 2.3 shows that the excitation system consists of three main components: the AVR, the DC exciter and the stabilising feedback. The AVR is modelled as a first-order transfer function, with regulator gain K_a and time constant T_a. A high gain value is desired in the AVR to achieve a rapid response and a low steady-state error. Despite a high value of AVR gain typically being set, this increases the likelihood of the exciter suffering from oscillatory instability in the power system. Therefore, a feedback loop is introduced into the excitation system to reduce the side effects of high AVR gain. The stabilising feedback loop is represented by a first-order differentiating function with a loop gain K_f and a time constant T_f. The DC exciter includes three components: an exciter gain K_e, a time constant T_e and a saturation function. The IEEE Type DC1A can be modelled by the set of differential equations presented in (2.14)–(2.16).

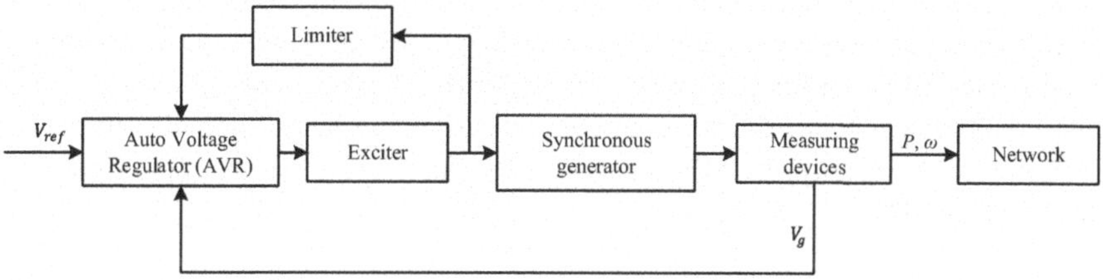

Fig. 2.2 The relationship between the SG, the excitation system and the network

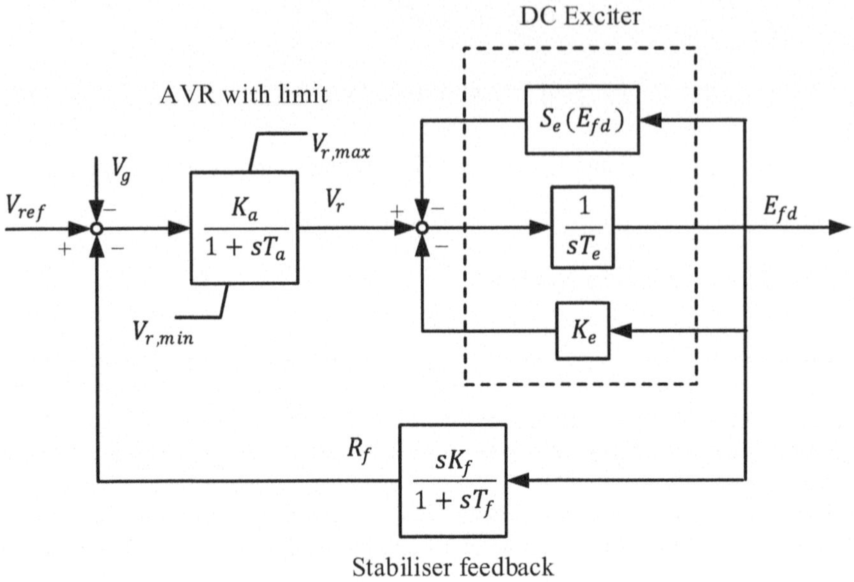

Fig. 2.3 Simplified block diagram for the IEEE Type DC1A exciter

$$T_e \frac{dE_{fd}}{dt} = V_r - (K_e + Se(E_{fd}))E_{fd} \quad (2.14)$$

$$T_a \frac{dV_r}{dt} = K_a R_f + K_a(V_{ref} - V_t) - V_r - \frac{K_a K_f}{T_f} E_{fd} \quad (2.15)$$

$$T_f \frac{dR_f}{dt} = -R_f + K_f / T_f K_e E_{fd} \quad (2.16)$$

In (2.14)–(2.16), $Se(E_{fd})$ represents the saturation function.

2.1.3 *The Overall SG Model*

Figure 2.2 illustrates the relationship between the SG and the excitation system. Overall, the excitation system modelled by (2.14)–(2.16) determines the field voltage based on the SG's terminal voltage. Subsequently, the SG, represented by (2.8)–(2.13), will then generate the current output according to the given field voltage. By combining (2.8)–(2.16), the SG's overall state-space model can be represented by Eqs. (2.17)–(2.18).

$$\frac{dx_i^{sg}}{dt} = A_i^{sg} x_i^{sg} + B_i^{sg} [v_{di,qi}^{sg}] \quad (2.17)$$

$$i^{sg}_{di,qi} = C^{sg}_i x^{sg}_i \tag{2.18}$$

In (2.17)–(2.18), x^{sg}_i represents the state variables of the ith SG; $v^{sg}_{di,qi}$ represents the SG terminal voltage in the dq frame; $i^{sg}_{di,qi}$ are the currents injected from the SG in the dq-frame.

In general, the SG is modelled as a voltage-controlled current source, with the input and output being the terminal voltage and the current, respectively.

2.1.4 The Branch Model

All transmission lines and transformers in this thesis are represented by a common branch model. The branch model shown in Fig. 2.4 consists of two components: a common π-equivalent transmission line model with series impedance $Z_{\text{line}} = R_{\text{line}} + jX_{\text{line}}$ and shunt susceptance B_c, and an ideal lossless transformer model with tap ratio magnitude γ and phase shift θ_{shift} [1, 8]. A standard π model with lumped parameter is adopted in this thesis for transmission line modelling with the assumption that transmission lines are short [9, 10]. The simplified transformer model with magnetising current neglected is connected at the sending end of the branch model, as shown in Fig. 2.4. It should be noted that when γ is set to 1, the branch model becomes an equivalent π transmission line model.

The relationship between the sending and receiving end voltages and currents can be expressed by the algebraic equations in (2.19), in which $Y_{\text{line}} = 1/Z_{\text{line}}$.

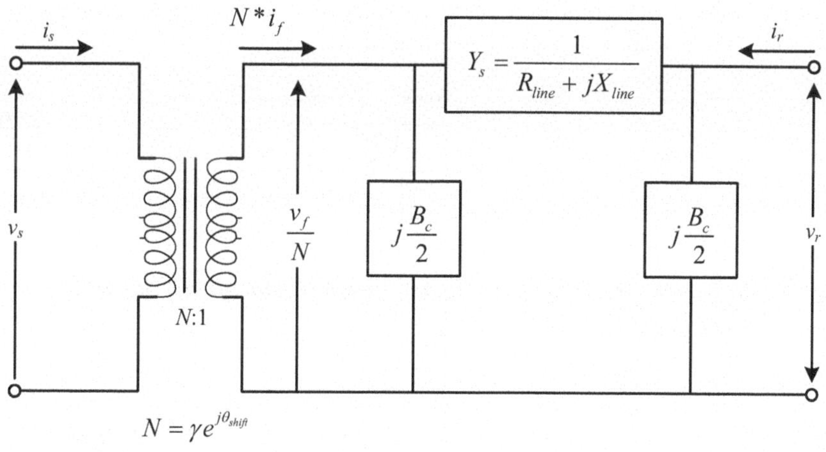

Fig. 2.4 Branch model

$$\begin{bmatrix} i_s \\ i_r \end{bmatrix} = \begin{bmatrix} \left(Y_{\text{line}} + j\frac{B_c}{2}\right)\frac{1}{\gamma^2} & -Y_{\text{line}}\frac{1}{\gamma e^{-j\theta_{\text{shift}}}} \\ -Y_{\text{line}}\frac{1}{\gamma e^{j\theta_{\text{shift}}}} & Y_{\text{line}} + j\frac{B_c}{2} \end{bmatrix} \begin{bmatrix} v_s \\ v_r \end{bmatrix} \tag{2.19}$$

In (2.19), i_s and i_r are the sending-end and receiving-end currents; v_s and v_r are the sending-end and receiving-end voltages.

2.1.5 The Load Model

The load represents all power-consuming devices connected to the power system, with its modelling having a significant impact on the system stability analysis [11]. The impact of different load representations on the system's small-disturbance voltage stability using a bifurcation theory-based analysis approach can be found in [12]. In this thesis, the constant power load is adopted and is represented as a shunt-connected admittance Y_i^{load} connected to the ith load, as in (2.20). Many previous studies, including [3, 5, 13], have adopted the constant power load model when using the local bifurcation theory-based method to investigate the system stability. Furthermore, the modelling of the load with a constant impedance is considered adequate for system stability analysis [14], particularly when undertaking small-disturbance stability analysis [3]. In analysis considering small-disturbance stability, the nonlinear power system is linearised and the use of a constant impedance representation has been typically used for load modelling. It should be noted that the modelling of the load is not critical for the analysis frameworks developed in this thesis, and that the impedance representation of the load is used in all the studies. More complicated and detailed load models, such as a nonlinear load model considering load recovery [15] or a time-variant dynamic load model [16], could be adopted to obtain a more accurate system dynamic response.

$$Y_i^{\text{load}} = \frac{P_i^{\text{load}} - jQ_i^{\text{load}}}{V_i^{\text{bus}^2}} \tag{2.20}$$

2.1.6 Network Modelling with Algebraic Equations (Static Network Model)

All of the electrical power system components are integrated via transmission lines and transformers. In the majority of past system stability research, particularly for large power networks, network oscillations between the series and shunt connected inductor and capacitor, i.e. LC dynamics, are neglected on the assumption that the power transfer speed through the network is considerably faster than the dynamics of the power system components. Hence, the network connecting all power system components is modelled with algebraic equations, with the network LC dynamics neglected. Accordingly, this is referred to as the static network model in this thesis. The static modelling of the network using algebraic equations results in a significant reduction in matrix dimensions, leading to a reduced computational burden. In this case, the system network consisting of all branches and loads can be represented as a set of nodal network equations. Considering a network with N buses, the nodal network matrix expressed in (2.21) represents the relationship between the busbar voltages, V, and currents, I, injected at all buses [14].

$$\begin{bmatrix} I_1 \\ \vdots \\ I_i \\ \vdots \\ I_N \end{bmatrix} = \begin{bmatrix} Y_{11} \dots Y_{1i} & & \dots Y_{1N} \\ \vdots & \ddots & \vdots & \ddots & \vdots \\ Y_{i1} & \dots & Y_{ii} & \dots & Y_{iN} \\ \vdots & \ddots & \vdots & \ddots & \vdots \\ Y_{N1} & \dots & Y_{Ni} & \dots & Y_{NN} \end{bmatrix} \begin{bmatrix} V \\ \vdots \\ V_i \\ \vdots \\ V_N \end{bmatrix} \tag{2.21}$$

In (2.21), I is the sum of the current injected at the ith node; V is the voltage at the ith node; Y_{ii} is the self-admittance at the ith node (which is the total admittance connected to the ith node); and Y_{ij} is the mutual admittance between the ith and jth nodes. The matrix Y is also known as the nodal admittance matrix. The dimension of matrix Y can be reduced using Kron reduction, which neglects all buses with zero current injection [17]. This results in a further reduction in the computational burden during power system analysis.

2.1.7 *Network Modelling with Ordinary Differential Equations (Dynamic Network Model)*

Recent research has evidenced the importance of considering network LC dynamics when assessing the stability of a power system, particularly for a power system in which multiple VSCs are connected [18, 19]. Unlike the conventional static network model, dynamic modelling of the AC network with first-order differential equations leads to a significant increase in the number of state variables in the system model. The AC network modelled with differential equations (considering the network LC dynamics) is referred to as the dynamic network model in this thesis. Considering a branch with an equivalent π representation, as shown in Fig. 2.4, the dynamic network model can be divided into two parts: the modelling of the transmission line and the modelling of the bus. The equations representing the dynamics of the transmission lines and buses are shown in (2.22)–(2.23).

$$L_{\text{line}} \frac{di_{\text{line}}}{dt} = v_s - v_r - R_{\text{line}} i_{\text{line}} \tag{2.22}$$

$$C\frac{\mathrm{d}v_{\mathrm{bus}}}{\mathrm{d}t} = \sum i \tag{2.23}$$

In (2.22)–(2.23), $\sum i$ represents the sum of the currents injected into the bus; v_{bus} is the bus voltage (i.e. it can represent either v_s or v_r in Fig. 2.4); i_{line} is the current passing through the transmission line; C is the sum of equivalent shunt capacitance connected to the bus.

All loads are modelled as series-connected impedances. The form of the dynamic equations for the loads will be the same as for the transmission lines presented in (2.22), with the receiving end voltage (v_r) equal to zero.

2.1.8 Network Reference Frames

The network equation in (2.21) is typically represented in phasor coordinates, with magnitude and angle. However, to reduce the complexity of the calculation, the SG and VSC are modelled with a *dq*-axis by an $abc - dq0$ transformation. Similarly, the system network is modelled in the global *DQ* domain to simplify AC network calculation. Each SG and VSC operates under its own local *dq*-frame, based on either its rotor angle or the angle provided by the phase-locked loop (PLL). Regarding the system reference frame, the reference angle from the selected reference machine is used as the basis of the network DQ frame in the model. In the ideal case, both the component's local *dq* frame and the network global *DQ* reference frame rotate at a synchronised speed ω_{syn} but are offset by a fixed angle. When multiple machines are included in the network, the current (and voltage) injected from each device must be transformed from the local *dq* frame to the global *DQ* frame. Equation (2.24) presents the transformation between the two coordinate systems. It should be noted that (2.24) can be applied to both voltage (V) and current (I).

$$\begin{pmatrix} v_d \\ v_q \end{pmatrix} = \begin{pmatrix} \cos(\alpha - \beta) & \sin(\alpha - \beta) \\ -\sin(\alpha - \beta) & \cos(\alpha - \beta) \end{pmatrix} \begin{pmatrix} v_D \\ v_Q \end{pmatrix}. \tag{2.24}$$

In (2.24), α is the reference angle of the global (DQ) frame, and β is the reference angle of the local (dq) frame.

2.2 The Voltage Source Converter

The main function of a VSC is to facilitate the exchange of energy between two subsystems (typically AC and DC subsystems) in a desired manner based on the performance specification. Figure 2.5 illustrates the main circuit of a modular multi-level VSC (MMC).

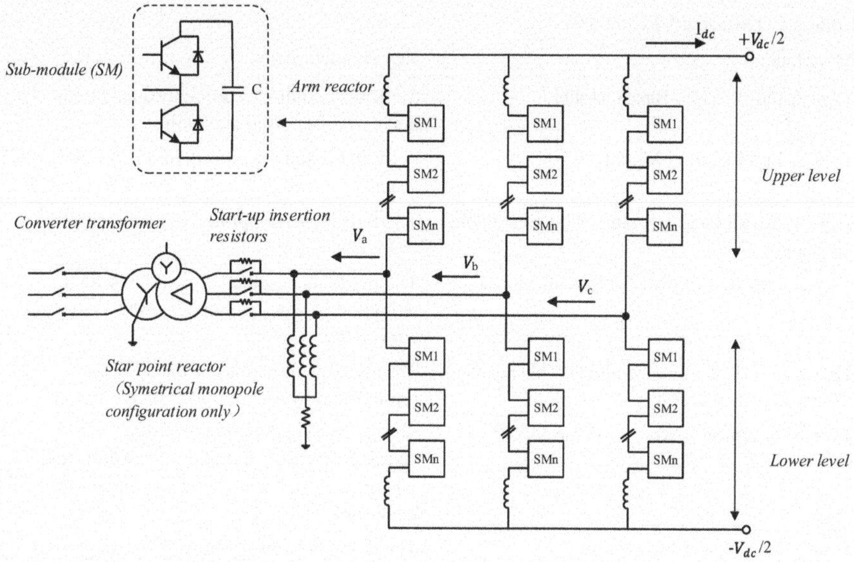

Fig. 2.5 Three-phase MMC

A typical MMC is powered by a DC voltage link and consists of a large number of sub-modules (SMs). Each SM is based on a half- or full-bridge cell, which operates as a switch. In terms of the modelling of a VSC, which is shown in Fig. 2.5, the SMs are located in two different sections, specifically the upper and lower levels, and as a first pass can be modelled with each level powered by a voltage source at a value of $V_{dc}/2$. At each level, SMs are connected in parallel and placed as three phases corresponding to the three phases of *abc* in the AC network. The SMs in each phase are integrated in series. Hence, by continuously regulating the operations of the SMs (i.e., the switching of each SM), the MMC can provide a wide range of voltage levels and generate an AC voltage waveform on the AC-side network.

In the past work, various VSC models with different levels of detail have been developed. Table 2.1 summarises seven VSC models from the CIGRE technical report [20], including their names and general information.

In this thesis, a Type-6 Simplified Average-Value Model will be used to represent the modelling of VSCs. Although studies with more detailed models (e.g. Type-4) will more accurately establish the system's dynamic response, it would require simulations of a large number of semiconductor switches with small time steps, which imposes a substantial computational burden. The research presented in [21] shows that the simplified AVM can effectively capture the essential dynamic performance when assessing the impact of the VSC on system stability. The AVM is a well-established VSC model that has been widely used in past research analysing VSC-related issues in accordance with small-disturbance stability [22–25]. Several impedance-based VSC models with different modelling details are developed in [26]. The work in [26] offers a systematic comparative study of VSC models with different

Table 2.1 Different VSC models

Model name	Key characteristics
Type 1. Full Physics Based Model	Switches and diodes are represented with detailed differential equation
Type 2. Full Detailed Model	A Model based on simplified IGBT (switches) models
Type 3. Model based on simplified switchable resistance	The IGBTs are simplified by resistors with changeable values
Type 4. Detailed Equivalent Circuit Model	The model is based on Type 3 but with a reduction of electrical nodes using Thevenin equivalents
Type 5. Average-Value Model (AVM) based on the switching function	The AC-side dynamics of the VSC are represented by a controlled voltage source
Type 6. Simplified Average-Value Model	Similar to Type 5, where AC-side characteristics are modelled as a controlled voltage source but with the switching function neglected
Type 7. RMS Load-Flow Model	The dynamics within the VSC are neglected and the VSC is modelled as a power source with constant output

frequency responses considering the power system instability caused by interactions between the VSC and the network. Moreover, with the application of modular multi-level VSCs, the AC output waveform is closer to the ideal sinusoid [20]. Consequently, the AC-side dynamic characteristics of the VSC can be represented by a controlled voltage source. More importantly, the frequency range of interest in the studies performed in this thesis is from 0.1 Hz to a few hundred Hz. Therefore, a switching function operating at 1 kHz works at a much higher bandwidth and hence has sufficient bandwidth separation from other power components. Therefore, the switching function is sufficiently fast to be neglected. Hence, the simplified AVM is considered adequate for a root-mean-square (RMS) level system stability analysis, and this VSC model will be used throughout the thesis. It should be noted that the analysis frameworks developed and the obtained analytical results in this thesis rely on comprehensive information from the VSC model, i.e., the CIGRE Type-6 VSC model.

Figure 2.6 illustrates the VSC model using a simplified AVM. The differential equation representing the AC-side dynamics of the VSC is provided by (2.25).

$$e_{abc} - v_{abc} = L\frac{\mathrm{d}i_{abc}}{\mathrm{d}t} + Ri_{abc} \tag{2.25}$$

In (2.25), i_{abc} are the currents flowing out of the VSC in the *abc*-frame; v_{abc} are the voltages at the PCC in the *abc*-frame; e_d, e_q are the VSC terminal voltage in the *dq*-frame and $R + j\omega L$ is the impedance of the VSC.

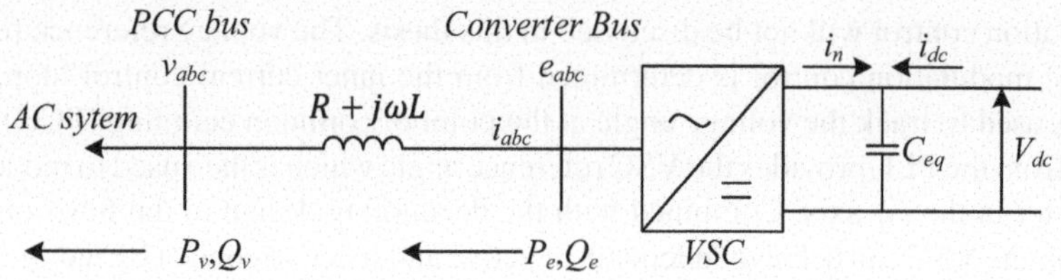

Fig. 2.6 VSC model using simplified AVM

With the *abc* – *dq* transformation based on the angle provided by the PLL, (2.25) becomes (2.26)–(2.27).

$$L\frac{\mathrm{d}i_d}{\mathrm{d}t} = e_d - v_d - Ri_d + \omega Li_q \tag{2.26}$$

$$L\frac{\mathrm{d}i_q}{\mathrm{d}t} = e_q - v_q - Ri_q - \omega Li_d \tag{2.27}$$

In (2.26)–(2.27), i_d, i_q are the currents flowing out of the VSC in the *dq*-frame; v_d, v_q are the voltages at the PCC in the *dq*-frame; e_d, e_q are the voltages at the terminal of the VSC in the *dq*-frame.

The DC-side dynamics of the VSC can be presented by (2.28)–(2.29).

$$C_{\mathrm{eq}}\frac{\mathrm{d}V_{\mathrm{dc}}}{\mathrm{d}t} = i_n + i_{\mathrm{dc}} \tag{2.28}$$

$$i_n V_{\mathrm{dc}} = e_d i_d + e_q i_q \tag{2.29}$$

In (2.28)–(2.29), C_{eq} is the VSC DC side equivalent capacitance; V_{dc} is the VSC DC-side voltage; i_n is the VSC output DC current; i_{dc} is the DC current from the DC network.

2.2.1 Overview of VSC Control

The vector control scheme is adopted for the studies in this thesis, as is discussed in Chap. 1. Figure 2.7 describes the generalised cascade control topology for the VSC used in this thesis based on vector control [27]. Generally, the control structure can be categorised into three layers: voltage modulation, inner current control and outer voltage control. The studies of the voltage modulation control, which represents the pulse width modulation (PWM) control for a two-level VSC and nearest level control for a modular multi-level VSC, are outside the scope of this thesis, given that the dynamic characteristics associated with the switching function are not critical in an RMS-level stability analysis. Therefore, the details of the switching-based voltage modulation control will not be discussed in this thesis. The voltage reference for the voltage modulation control is determined from the inner current control loop. The PLL is used to track the voltage angle at the point of common coupling (PCC). The angle from the PLL provides the VSC reference angle which is then used to model the converter in the *dq*-axis. Combined with the decoupling design in the inner current control, the VSC can achieve independent active and reactive control by individually regulating the *dq* current components. The *d*-axis current component determines the VSC active power output, with the reference value provided by either the outer power loop or the DC voltage control. The VSC reactive power is manipulated by controlling the *q*-axis current component, with the reference value provided either by the outer reactive power loop or the AC voltage control.

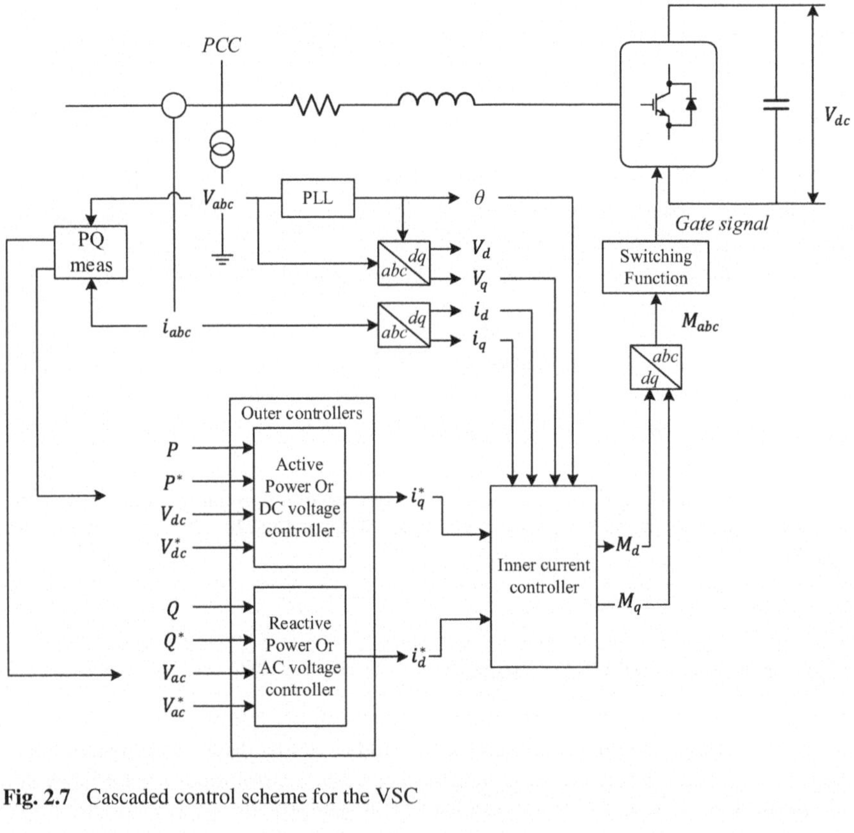

Fig. 2.7 Cascaded control scheme for the VSC

2.2.2 Voltage Modulation Control

As discussed previously in Sect. 2.2.1, the voltage modulation control (the switching function) in the VSC will not be discussed in detail, as it is outside the scope of this thesis. In this thesis, the voltage modulation control is simplified and modelled as a first-order transfer function and given by (2.30)–(2.31).

$$\tau \frac{\mathrm{d}e_d}{\mathrm{d}t} = (e_d^* - e_d) \tag{2.30}$$

$$\tau \frac{\mathrm{d}e_q}{\mathrm{d}t} = (e_q^* - e_q) \tag{2.31}$$

In (2.30)–(2.31), τ is the time constant, which is set to 82 μs in this thesis based on [28]; e_d^*, e_q^* are the VSC reference terminal voltages in the *dq*-frame.

2.2.3 Vector Current Control

Vector control (also known as *dq* current control) is implemented by converting the current and voltage into the *dq* domain based on a reference angle provided by the PLL [27, 29]. Figure 2.8 illustrates the control scheme of the VSC in *dq* frame with outer *PQ* feedback loop control.

With the PLL tracking the voltage angle from the PCC, the *dq* frame rotates at the system synchronous speed in a balanced steady-state condition. With this PLL configuration, the d-axis component is aligned with the phase-A voltage vector on the PCC bus, and hence the time-variant voltage is represented by a time-invariant DC value. As evidenced in (2.26)–(2.27), which represent the VSC AC-side dynamics, the d-axis and q-axis current components are coupled through the terms $\omega L i_q$ and $\omega L i_d$. The *dq*-axis current control loop is decoupled using the nulling terms to reduce the cross-coupling effect, as shown in Fig. 2.8. With such decoupling techniques, combined with *dq* modelling, the active and reactive power output of the VSC can be regulated independently by individually manipulating the d-axis and q-axis current components [30]. The dynamics of the inner current loop are represented in the *dq*-axis, as shown in (2.32)–(2.35).

$$e_d^* = K_{p_{id}}\left(i_d^* - i_d\right) + x_d - \omega L i_q + v_d \tag{2.32}$$

$$e_q^* = K_{p_{iq}}\left(i_q^* - i_q\right) + x_q + \omega L i_d + v_q \tag{2.33}$$

$$\frac{\mathrm{d}x_{id}}{\mathrm{d}t} = K_{i_{id}}\left(i_d^* - i_d\right) \tag{2.34}$$

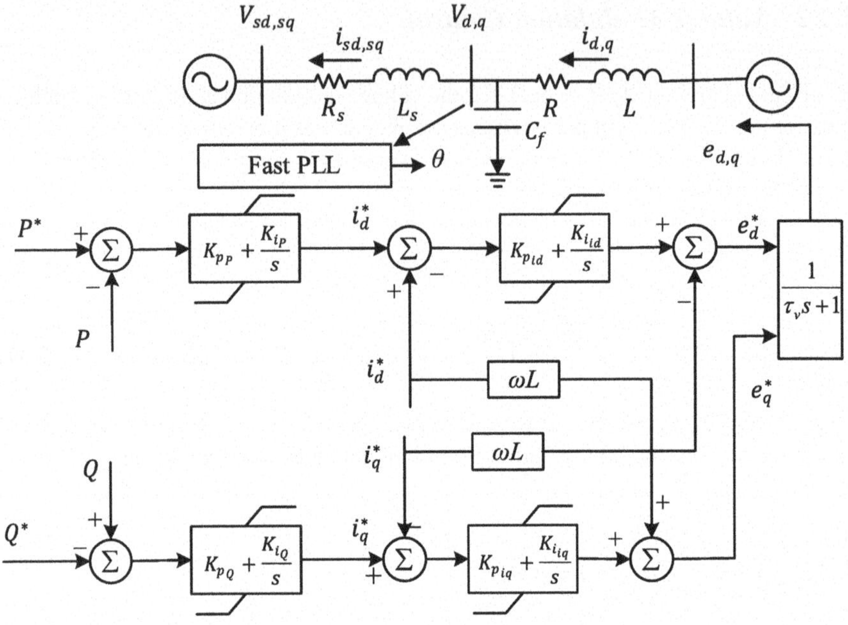

Fig. 2.8 Control scheme of the VSC in *dq* frame with outer PQ feedback loop control

$$\frac{dx_{iq}}{dt} = K_{i_{iq}}\left(i_q^* - i_q\right). \tag{2.35}$$

In (2.32)–(2.35), $K_{p_{id,q}}, K_{i_{id,q}}$ are the controller gains of the inner current PI controllers in the *dq*-axis, respectively; i_d^*, i_q^* are the set reference values of the current across the VSC reactor in the *dq*-axis, which are determined by the outer loop; and x_d, x_q are the state variables of the integrators of the current loops in the *dq*-axis respectively.

The transfer function of the inner current control can be developed from the closed-loop block diagram shown in Fig. 2.8. Subsequently, the developed transfer function then can be used to tune the PI control parameters of the inner current control using the Skogestad internal model control (SIMC) tuning method described in [31].

2.2.4 The Phase-Locked Loop

The PLL is the fundamental component of the vector control, as it provides the reference angle for the $abc - dq$ transformation. The PLL's performance directly affects the system's overall dynamic response, particularly when assessing the interaction between VSC control loops. Hence, the PLL modelling is critical for the power system stability analysis.

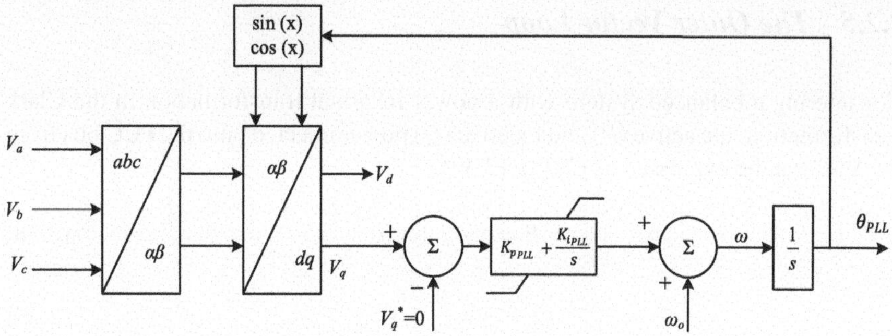

Fig. 2.9 Block diagram of the SRF PLL

The synchronous reference frame (SRF) PLL presented in the block diagram in Fig. 2.9 is the most widely used three-phase PLL technique in power system research [29, 32, 33].

As mentioned in Sect. 2.1, this thesis analyses power stability issues at the transmission system level, at which all three phases are assumed to be equal and balanced. Hence, SRF-PLL is adopted in studies carried out in this thesis. In terms of the application of the PLL, the instantaneous three-phase voltage (v_{abc}) at the PCC will be measured and converted into DC quantities in the *dq* frame (v_{dq}) based on the angle from the PLL to form a feedback loop. Under balanced steady-state conditions, the PLL is configured to ensure the voltage at the PCC bus is aligned with the *d*-axis in the *dq* frame. Hence, the PLL controller, typically a PI controller, is used to control the frequency of the PLL by driving the *q*-axis voltage to zero [28]. Following this, the angle used for $abc - dq$ transformation is then calculated by integrating the frequency from the PI controller in the PLL. The corresponding dynamics of the PLL are represented in the *dq* frame, as expressed in (2.36)–(2.37).

$$\frac{\mathrm{d}\theta_{\mathrm{PLL}}}{\mathrm{d}t} = K_{p_{\mathrm{PLL}}}(v_q - v_q^*) + x_{\mathrm{PLL}} \tag{2.36}$$

$$\frac{\mathrm{d}x_{\mathrm{PLL}}}{\mathrm{d}t} = K_{i_{\mathrm{PLL}}}(v_q - v_q^*) \tag{2.37}$$

In (2.36)–(2.37), θ_{PLL} is the angle measured by the PLL; $K_{p_{\mathrm{PLL}}}$, $K_{i_{\mathrm{PLL}}}$ are the PI gains of the PLL controller; x_{PLL} is the state variable of the integrator in the PLL controller; v_q^* is the reference *q*-axis voltage, which is typically set to zero.

Similar to the inner current loop, the SIMC tuning method can be applied to the PLL based on the transfer function developed in Fig. 2.9.

2.2.5 The Outer Vector Loop

Considering a balanced system with a power invariant transformation in the Clark transformation, the active (P) and reactive (Q) power injected onto the PCC bus from the VSC can be expressed as (2.38)–(2.39).

$$P = v_d i_d + v_q i_q \tag{2.38}$$

$$Q = -v_d i_q + v_q i_d \tag{2.39}$$

The outer loop controls can be categorised into two types: feedback control and feedforward control. Feedback control involves calculating the error between the measured and reference values. This error is fed into a PI controller to generate current references for the inner loop and hence form a feedback loop. The feedforward control approach, rather than incorporating an additional PI controller into the system, directly calculates the current set point value based on (2.38)–(2.39), assuming that $v_q = 0$ (i.e., that the PLL works perfectly) [34]. Although the feedforward approach offers a faster response speed, it makes a number of assumptions which may not always be strictly true in real powers and therefore potentially increases instability, particularly when the PLL is inaccurate during disturbances. This thesis adopts feedback controls for the VSC outer loop, because these are the most widely used control topologies in past power systems research [27, 35, 36]. More in-depth comparisons of outer feedback and feedforward loop controls can be found in [37].

2.2.5.1 Active and Reactive Power Control

The VSC power output can be maintained by introducing two additional feedback outer controls that separately regulate the *dq*-axis current reference, as shown in Fig. 2.8. The dynamics related to the outer loop for the control of active and reactive power loops can be represented in (2.40)–(2.43).

$$i_d^* = K_{p_P}(P^* - P) + x_P \tag{2.40}$$

$$i_q^* = K_{p_Q}(Q - Q^*) + x_Q \tag{2.41}$$

$$\frac{\mathrm{d}x_P}{\mathrm{d}t} = K_{i_P}(P^* - P) \tag{2.42}$$

$$\frac{\mathrm{d}x_Q}{\mathrm{d}t} = K_{i_Q}(Q - Q^*) \tag{2.43}$$

In (2.40)–(2.43), i_d^*, i_q^* are the reference currents in the dq components calculated by the outer loop, and will be fed into the inner current loop; $K_{p_{P,Q}}$, $K_{i_{P,Q}}$ are the proportional and integral gains of the outer PI-based PQ controllers in the dq-axis respectively; P^* and Q^* are the power reference for the PQ outer controller. P and Q are the power injected into the PCC bus of the VSC.

2.2.5.2 DC Voltage Control

Large variations in the DC voltage suggest that the power transfer between the AC and DC networks is unbalanced, which is not permissible. Hence, the DC-voltage outer loop is employed to keep the DC voltage link constant. In a real power network with multiple VSCs connected to the same DC network, at least one of the VSCs should be responsible for maintaining a constant DC voltage. This is achieved by incorporating a PI controller that modifies the d-axis current with the input of the measured DC link voltage. The differential equations describing the outer DC voltage control are provided by (2.44)–(2.45).

$$i_d^* = K_{p_{V_{\mathrm{dc}}}}\left(V_{\mathrm{dc}}^* - V_{\mathrm{dc}}\right) + x_{V_{\mathrm{dc}}} \tag{2.44}$$

$$\frac{\mathrm{d}V_{\mathrm{dc}}}{\mathrm{d}t} = K_{i_{V_{\mathrm{dc}}}}\left(V_{\mathrm{dc}}^* - V_{\mathrm{dc}}\right) \tag{2.45}$$

In (2.44)–(2.45), $K_{p_{V_{\mathrm{dc}}}}$, $K_{i_{V_{\mathrm{dc}}}}$ are the proportional and integral gains of the outer DC voltage controllers in the dq-axis respectively; V_{dc}^* is the DC voltage reference for the DC voltage outer controller.

2.2.5.3 AC Voltage Control

The VSC can control the AC voltage at the PCC, providing AC-side voltage support by applying the outer AC voltage control. The AC voltage controller drives the voltage sampled from the PCC bus to reach its set reference point by regulating the q-axis current reference. Typically, the AC-voltage outer loop is preferred when a weak AC grid is connected. The dynamics of the AC-voltage outer control are represented in (2.46)–(2.48).

$$i_q^* = K_{p_{V_{\mathrm{ac}}}}\left(V_{\mathrm{ac}}^* - V_{\mathrm{ac}}\right) + x_{V_{\mathrm{ac}}} \tag{2.46}$$

$$\frac{\mathrm{d}x_{V_{\mathrm{ac}}}}{\mathrm{d}t} = K_{i_{V_{\mathrm{ac}}}}\left(V_{\mathrm{ac}}^* - V_{\mathrm{ac}}\right) \tag{2.47}$$

$$V_{\mathrm{ac}} = \sqrt{v_d^2 + v_q^2} \tag{2.48}$$

In (2.46)–(2.48), $K_{pv_{ac}}$, $K_{iv_{ac}}$ are the proportional and integral gains of the outer AC voltage controllers in the dq-axis respectively; V_{ac}^* is the AC voltage reference for the AC voltage outer controller; V_{ac} is the AC voltage at the PCC bus.

2.2.6 The Overall VSC Model

Based on the overall cascaded control topologies shown in (2.38)–(2.48) and combining (2.26)–(2.37), each VSC, including the dynamics of the PLL, can be constructed in the state-space form. The complete state-space form of the ith VSC (if the PQ outer loop is assumed to have been installed) is shown in (2.49)–(2.50).

$$\frac{dx_i^{vsc}}{dt} = A_i^{vsc} x_i^{vsc} + B_i^{vsc}[P_i^*, Q_i^*, v_{di,qi}^{vsc}] \tag{2.49}$$

$$i_{di,qi}^{vsc} = C_i^{vsc} x_i^{vsc} \tag{2.50}$$

In (2.49)–(2.50),x_i^{vsc} represents the state variables of the ith VSC; $v_{di,qi}^{vsc}$ represents the VSC terminal voltage in dq frame; $i_{di,qi}^{vsc}$ are the currents injected from the VSC in dq-frame.

Similar to the SG, the VSC is represented as a voltage-controlled current source in this thesis.

The study within this thesis focuses on analysing dynamic response of power systems with multiple VSCs coupled through AC networks and the impacts of these VSC-interfaced devices on the AC-side dynamics of the system. In general, except for the most significant deviations in the DC network, the voltage modulation function implemented in conventional VSCs acts as a feed-forward compensatory mechanism for variations in DC link voltage. Additionally, the work in [38] has demonstrated that the AC and DC sides are de-coupled by the VSC when analysing the system's small-disturbance stability if the VSC does not regulate its DC voltage output. Hence, the DC-side dynamics will not propagate through to the AC network and the power transfers within the VSC, as considered in this thesis, inherently remain independent of the dynamics governing the DC-side network. Hence, only the AC-side dynamics are modelled in the thesis, in accordance with the assumption that the VSC is connected to a stiff DC network in which DC-side dynamics are perfectly controlled by the other end of the VSC.

To further demonstrate the validity of the assumption of neglecting the DC side dynamics, Appendix C includes a comparative study of dynamic responses of the system, using different DC network representations. It has been clearly demonstrated that a VSC-HVDC half-link model, where a standalone VSC is directly connected to a stiff DC power source has a similar performance with the VSC-HVDC full-link model, where the entire DC link is included, during a small-disturbance event. Consequently, this thesis adopts the VSC-HVDC half-link model. Many previous studies, including [23, 25, 39], have adopted this modelling convention when investigating interactions between VSCs that are coupled through the AC network.

2.3 Overall System Integration

Once each power component model has been constructed, the subsequent crucial stage involves integrating all power system components to formulate an overall model to perform the system stability analysis. The following section illustrates power system integration via a network modelled by algebraic equations (static network model) and represented by differential equations (dynamic network model), respectively.

2.3.1 System Integration with the Static Network Model

Figure 2.10 shows the schematic diagram of the completed system model with a network modelled using algebraic equations. As Fig. 2.10 presents, the output (current) of the SGs and VSCs represented in (2.17)–(2.18) and (2.49)–(2.50) match the injection currents in the left-hand side of (2.21). The right-hand side of (2.21) provides the voltages which are the input to the VSCs and SGs in (2.17)–(2.18) and (2.49)–(2.50). In general, the output currents from the VSCs and SGs are fed into the network, and the network then feeds back the voltages.

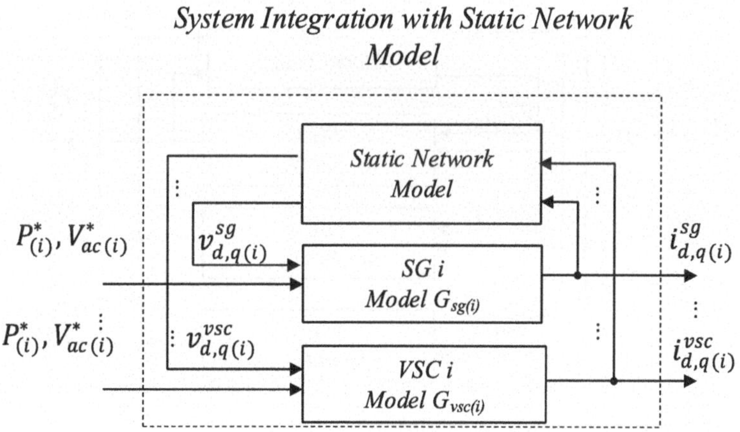

Fig. 2.10 System integration with the static network model

2.3.2 System Integration with the Dynamic Network Model

Figure 2.11 shows the integration of power system components using the dynamic network model. Generally, the dynamics of a busbar work as a junction that interconnects other AC components, including AC transmission lines, loads, VSCs and SGs. The busbar model's input (2.23) denotes the current injections from other components (i.e. the output of the SG, VSC and transmission line models in (2.17)–(2.18), (2.49)–(2.50) and (2.22)). Whilst the output of the busbar model is the calculated voltages, which are the input to the VSC, SG and transmission line models. With the busbar differential equations, all components on the AC side of the network are integrated together to form a closed loop.

2.4 Power System Analysis Techniques

The previous section described how the power system will be modelled based on the mathematical equations of the various relevant power system components. This section will introduce the linear (modal analysis) and nonlinear (bifurcation theory) analysis tools that are adopted in this thesis to assess the small-disturbance stability of the power system.

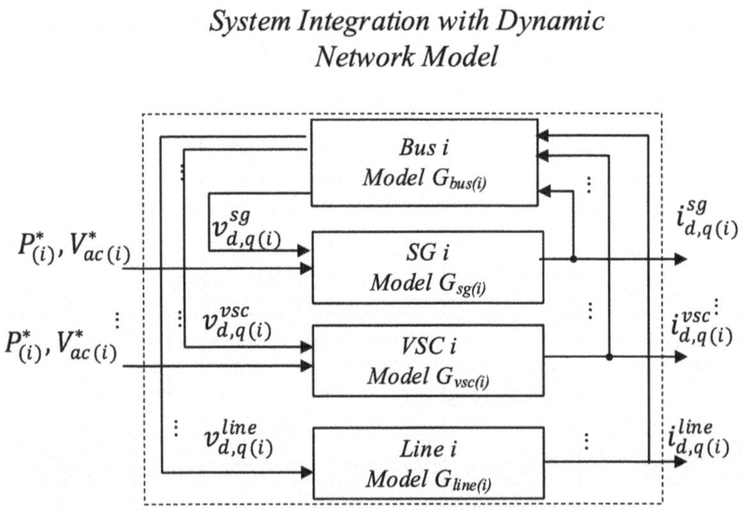

Fig. 2.11 System integration with the dynamic network model

2.4.1 Modal Analysis

As discussed previously in Chap. 1, small-disturbance stability represents the system's ability to remain in stable operation after the injection of small disturbances [1]. As the magnitude of the disturbances considering small-disturbance stability is considered sufficiently small, the power system can be linearised around the steady-state operating point for the purpose of analysis. The dynamics of power systems can be described using sets of differential–algebraic equations (DAEs), as shown in (2.51).

$$\begin{cases} \frac{\mathrm{d}x}{\mathrm{d}t} = f(x, y, \mu) \\ 0 = g(x, y, \mu) \end{cases} \tag{2.51}$$

A system equilibrium point (x_0, y_0, μ_0) can be defined by equating the left-hand side of (2.51) to zero. Considering a small perturbation subjected to this point, the new states are presented as $(x_0 + \Delta x, y_0 + \Delta y, \mu_0 + \Delta\mu)$, where Δ denotes a small deviation. Consequently, (2.52) is established.

$$\begin{cases} \frac{\mathrm{d}x_0}{\mathrm{d}t} + \frac{\mathrm{d}\Delta x}{\mathrm{d}t} = f(x_0 + \Delta x, y_0 + \Delta y, \mu_0 + \Delta\mu) \\ 0 = g(x_0 + \Delta x, y_0 + \Delta y, \mu_0 + \Delta\mu) \end{cases} \tag{2.52}$$

The first-order Taylor series expansion of (2.52) is deemed to be a sufficient approximation of system dynamic behaviour, given that only small disturbances are considered [1]. The linearised state-space power system model is expressed in (2.53).

$$\begin{bmatrix} \frac{\mathrm{d}\Delta x}{\mathrm{d}t} \\ 0 \end{bmatrix} = \begin{bmatrix} f_x & f_y \\ g_x & g_y \end{bmatrix} \begin{bmatrix} \Delta x \\ \Delta y \end{bmatrix} \tag{2.53}$$

In (2.53), f_x, f_y, g_x, g_y denote the partial derivation of functions f and g with respect to x and y.

If g_y in (2.53) is not singular (i.e. $\det(g_y) \neq 0$), then Δy can be represented by a function of Δx, as indicated in (2.54).

$$\Delta y = -g_y^{-1} g_x \Delta x \tag{2.54}$$

Substituting (2.54) into (2.53) results in (2.55)–(2.56).

$$\frac{\mathrm{d}\Delta x}{\mathrm{d}t} = A_{\mathrm{sys}} \Delta x \tag{2.55}$$

$$A_{\mathrm{sys}} = f_x - f_y g_y^{-1} g_x \tag{2.56}$$

A_{sys} is referred to as the dynamic system state matrix. Based on Lyapunov's first method, the roots of the characteristic equation of the system's first approximation determine the small-disturbance stability of the system. For a linearised state-space power system model, the fundamental small-disturbance dynamic characteristics of the model are expressed in terms of the eigenvalue properties of the reduced system matrix A_{sys}.

Assuming that A_{sys} is an $n \times n$ matrix, the eigenvalues of A_{sys} are determined by the values of the scalar λ, for which there are non-trivial solutions (i.e. $\phi \neq 0$) to (2.57).

$$A_{sys}\phi = \lambda\phi \tag{2.57}$$

To calculate eigenvalues, (2.57) can be written as (2.58).

$$(A_{sys} - \lambda I) = 0 \tag{2.58}$$

Considering the solutions are non-trivial, the eigenvalues can be obtained using (2.59).

$$\det(A_{sys} - \lambda I) = 0 \tag{2.59}$$

The n solutions of (2.59) are the n eigenvalues ($\lambda = \lambda_1 \ldots \lambda_n$) of matrix A_{sys}.

For any given eigenvalue, the $n \times 1$ column vector which satisfies (2.57) is named as the right eigenvector of A_{sys} associated with λ_i. Similarly, the $1 \times n$ row vector, which satisfies (2.60), is called the left eigenvector of A_{sys} associated with eigenvalue λ_i.

$$\psi A_{sys} = \psi\lambda \tag{2.60}$$

The left and right eigenvectors are orthogonal when corresponding to different eigenvalues (i.e., $\psi_j\phi_i = 0$). Moreover, the eigenvectors are commonly normalised so that $\psi_i\phi_i = 1$, as neither has units.

Each pair of eigenvalues ($\lambda_i = \sigma_i + j\omega_i$) represents one oscillatory mode in the power system [9]. The time-based behaviour of each eigenvalue pair is provided by $e^{\lambda_i t}$ [14]. As shown in the time-domain expression (2.61), the real part of eigenvalue (σ) represents the damping of the mode, whilst the imaginary part (ω) represents the mode's oscillatory frequency. The system is considered small-disturbance stable only when all eigenvalues of the system have negative real part.

$$\lambda = \sigma \pm j\omega \tag{2.61}$$

The right eigenvector, also known as the mode shape, offers information about the relative activity of the state variable when a system mode is excited. The left eigenvector indicates the influence of the state variable in the associated mode. Thus, the kth element of the right eigenvector measures the involvement of the state variable

x_k in the ith mode, and the kth element of the left eigenvector weights the contribution of this involvement to the ith mode. A new index called the participation factor combines the right and left eigenvectors and hence provides a measure of the associated degree of the mode and state variables [1]. The calculation of the participation factor is expressed in (2.62).

$$P_i = \begin{bmatrix} p_{1i} \\ p_{2i} \\ \vdots \\ p_{ni} \end{bmatrix} = \begin{bmatrix} \phi_{1i}\psi_{1i} \\ \phi_{2i}\psi_{2i} \\ \vdots \\ \phi_{ni}\psi_{ni} \end{bmatrix} \tag{2.62}$$

The element p_{ki} is a measure of the relative participation of kth state variable in the ith mode. If the participation factor has a large magnitude, this indicates a significant degree of participation of the state variable in the relevant mode.

2.4.2 Bifurcation Theory

Apart from linear analysis tools based on the linearised system model, there are nonlinear analysis approaches such as bifurcation theory. In a nonlinear dynamic system, the continuous change of a system's parameter can potentially lead to a critical point being reached at which a qualitative change occurs in the system's topological structure. This critical point is generally referred to as a type of bifurcation point, which can be identified by applying bifurcation theory [3, 40].

A power system contains an abundance of nonlinear components, such as generators, switching power converters and protective devices. Hence, a power system is typically represented by sets of nonlinear DAEs, as shown in (2.51), in which μ is a vector of time-invariant scalar parameters and is also known as the bifurcation parameter.

Similar to (2.54), in regions in which g^{-1} is not singular, the algebraic variable y can be expressed as a function of the state variable x, as indicated in (2.63).

$$y = g^{-1}(x, \mu) \tag{2.63}$$

Substituting (2.63) into (2.51) results in (2.64). Consequently, the original DAE's represented power system model can be simplified as ordinary differential equations (ODEs).

$$\frac{dx}{dt} = f\left(x, g^{-1}(x, \mu), \mu\right) = F(x, \mu) \tag{2.64}$$

The system equilibrium point (also known as the fixed point) can be defined by equating the left-hand term $\dot{x}$ of (2.64) to zero, as in (2.65).

$$0 = F(x, \mu) \tag{2.65}$$

Supposing (x_0, μ_0) is the system equilibrium point, then (2.65) specifies the location of the equilibrium points. If the eigenvalue of the Jacobian becomes non-zero (i.e., F is invertible), then, based on the implicit function theorem, (2.65) can be transformed to (2.66) [3]. Equation (2.66) shows that the location of the system equilibrium point x_0 can be expressed as a smooth function of the bifurcation parameter μ_0. Consequently, with μ_0 varying, x_0 changes correspondingly in the state space, determining the system's equilibrium trajectory.

$$x = x(\mu) \tag{2.66}$$

The small-disturbance stability of the fixed point along the system's equilibrium trajectory can be determined by the eigenvalue analysis of A_{sys} from (2.56). If the power system is represented as sets of ODEs, as in (2.64), the system Jacobian matrix $\partial F/\partial x$ is the dynamic system state matrix, A_{sys}. When the eigenvalues of the system Jacobian matrix have negative real parts, the equilibrium point obtained from (2.65) is asymptotically stable. At the critical point, μ_c, the system eigenvalues will pass the imaginary axis, several system equilibrium branches given by (2.66) may come together and the system loses its structural stability. Thus, μ_c is then defined as the bifurcation value, and (x_c, μ_c) is referred as the bifurcation point.

2.4.2.1 Different Types of Bifurcations

Local bifurcation theory has been used to analyse system stability issues in power systems. Four types of bifurcations are related to power system instability and have been widely investigated in previous research: Saddle-node bifurcation (SNB), Hopf bifurcation (HB), Limit induced Bifurcation (LIB) and Singular induced Bifurcation (SIB) [5, 41, 42].

(1) An SNB occurs when two system equilibrium points, one stable and another unstable, merge, and the system equilibrium point disappears with the change of the bifurcation parameter. At the SNB, the system Jacobian matrix becomes singular, and a real eigenvalue becomes zero [43]. The condition for determining the SNB can be summarised as (2.67)–(2.69) [42, 44, 45].

$$F(x_c, \lambda_c, p) = 0 \tag{2.67}$$

$$\text{eig}\left(\frac{\mathrm{d}F}{\mathrm{d}x_c}\right) = 0 \tag{2.68}$$

$$d\left(\text{Re}\left(\text{eig}\left(\frac{\mathrm{d}F}{\mathrm{d}x_c}\right)\right)\right) \neq 0 \tag{2.69}$$

(2) An HB occurs when a stable system equilibrium point expands into a limit cycle or vice versa. If the newly born limit cycle is stable, it is called supercritical Hopf bifurcation. In the supercritical case, the limit cycle is an attractor, and the system's operational state is attracted by this limit cycle. Otherwise, if the limit cycle is unstable, it is called the subcritical Hopf bifurcation, and the newly born limit cycle is a repeller, which drives the system state away. The appearance of the HB is accompanied by a pair of complex conjugate eigenvalues passing the imaginary axis to or from an unstable zone [43]. Mathematically, the conditions for determining the HB are shown in (2.70)–(2.72) [42, 44, 45]:

$$(x_c, \lambda_c, p) = 0 \tag{2.70}$$

$$\mathrm{eig}\left(\frac{\mathrm{d}F}{\mathrm{d}x_c}\right) = \pm i\omega \tag{2.71}$$

$$d\left(\mathrm{Re}\left(\mathrm{eig}\left(\frac{\mathrm{d}F}{\mathrm{d}x_c}\right)\right)\right) \neq 0 \tag{2.72}$$

(3) A LIB occurs when the system state jumps from one to another because of the limitations of the system state variables. In a power system, with the increasing bifurcation parameter (which is commonly selected as the system loading), certain system state variables may reach their limits. As a result, the previous state variables are replaced by constant values, which leads to a change in the topological structure of the system. Hence, as the bifurcation parameters are regulated continuously, the states of the system discontinuously vary from one to another when the LIB is encountered. The stability of the new equilibrium point after the LIB is given by the eigenvalue analysis of the new system dynamic state matrix. Further investigations of the LIB (along with the SNB and HB) are performed in Chap. 3.

(4) A SIB occurs when system equilibrium is singular. When the SIB appears, some of the system eigenvalues may become unbounded. The investigation into the SIB is outside the scope of this thesis, as there is a lack of system instabilities in real power systems caused by SIBs. Details of the SIB are discussed in [5, 46].

In general, the SNB and LIB are considered as static bifurcations since they represent structural changes in the system's equilibrium curve. The HB is classified as a dynamic bifurcation which can be analysed using modal analysis. The small-disturbance stability discussed in Sect. 1.1 is commonly associated with the HB. The power system issues in accordance with voltage stability can be identified by considering all three types of bifurcations, i.e. the SNB, HB and LIB.

2.4.2.2 The Continuation Method

Consider a power system represented by (2.64), in which μ is the system loading. In general, as the system loading continuously increases, the bus voltage will gradually reduce, which forms a typical nose curve, shown in Fig. 2.12.

One of the applications of bifurcation theory in power systems is to determine the steady-state stability limit from the nose curve, which is the system equilibrium curve when the system loading is chosen as the bifurcation parameter. One direct way of obtaining the system equilibrium curve is by solving (2.65) with different bifurcation parameter values. However, the system Jacobian matrix $(\partial F/\partial x)$ becomes singular at the maximum power point (i.e., the nose tip of the nose curve) [3]. This results in a convergence issue if the traditional Newton–Raphson method is applied to solve (2.65). In this thesis, computational techniques based on the continuation and bifurcation methods are employed to avoid the convergence problem. The continuation method is used in this thesis to trace the system equilibrium trajectory from the initial point with constantly changing bifurcation parameters. Figure 2.12 illustrates the application progress of the continuation approach. Generally, the continuation method involves two steps: the predictor and the corrector [3]. It should be noted that a proper parameterisation technique should be selected before the predictor and corrector steps in order to initialise the system appropriately.

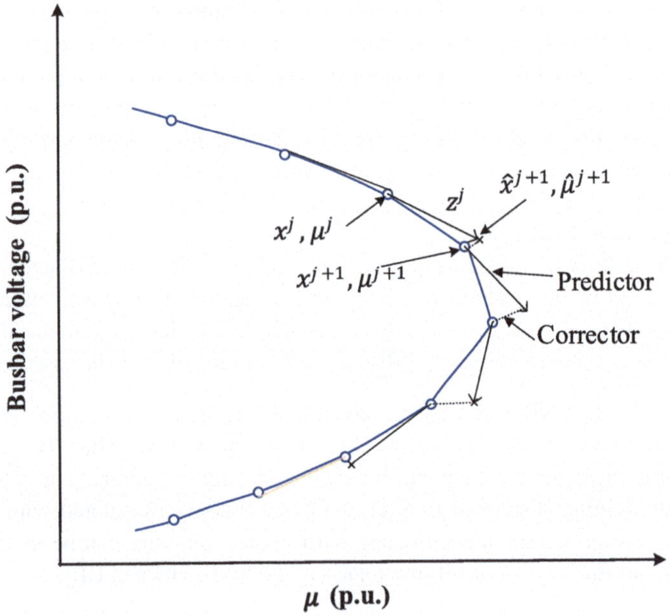

Fig. 2.12 Continuation method

Parameterisation

Assuming that the system represented by (2.64) has n state variables, there are $n + 1$ variables in (2.65) as the additional bifurcation parameter is introduced. To solve (2.65), an additional equation φ called the parametrisation equation, is introduced. Parameterisation can be viewed as a mathematical way of quantifying each solution so that the next and previous solutions are interrelated and quantified. Several techniques have been developed for the parameterisation scheme.

Natural parameterisation directly uses μ as the parameter [8, 47]. Hence, the μ^j in the next solution is simply the previous value plus the step size, which is presented in (2.73).

$$\varphi^j(x, \mu) = \mu - \mu^j - o^j \tag{2.73}$$

In (2.73), o^j is the step size of the current prediction.

However, as mentioned previously, the system's Jacobian matrix will become singular at some critical point (e.g. the maximum loading point). At this point, the system suffers from a convergence problem if natural parameterisation is used. This means that the natural parameterisation has become ill-suited for parameterising the curve. Pseudo-arc-length parameterisation is used to overcome this problem. In the pseudo-arc-length parameterisation scheme, the next point on the equilibrium curve is constrained to lie in the hyperplane running through the prediction point $(\widehat{x}^{j+1}, \widehat{\mu}^{j+1})$ orthogonal to the tangent line from the current point (x^j, μ^j) on the curve [48]. The relationship between the previous and next solutions defined by the pseudo-arc-length parameterisation scheme is given by (2.74).

$$\varphi^j(x, \mu) = \left(\begin{bmatrix} x \\ \mu \end{bmatrix} - \begin{bmatrix} x^j \\ \mu^j \end{bmatrix} \right)^T \overline{z^j} - o^j \tag{2.74}$$

The Predictor

The predictor aims to predict the next fixed point on the system's equilibrium curve based on the current equilibrium point. The first step is to determine the changing direction of the system state variables by calculating the vector that is tangent to the current equilibrium curve. Considering a system represented by (2.64), the tangent vector $z^j = [\mathrm{d}x^j \ \mathrm{d}\mu^j]$, is obtained by (2.75).

$$z^j = [D_x F|_{x^j, \mu^j}]^{-1} \frac{\mathrm{d}F}{\mathrm{d}\mu}|_{x^j, \mu^j} \tag{2.75}$$

Then, the obtained tangent vector is normalised, as presented in (2.76).

$$\overline{z^j} = \frac{z^j}{\|z^j\|} \tag{2.76}$$

Following this, the prediction of the next equilibrium point is calculated in (2.77).

$$\begin{bmatrix} \hat{x}^{j+1} \\ \hat{\mu}^{j+1} \end{bmatrix} = \begin{bmatrix} x^j \\ \mu^j \end{bmatrix} + o^j z^j \tag{2.77}$$

The Corrector

The purpose of the corrector is to use the approximation provided by the predictor to identify the real solution of the subsequent equilibrium point. Newton–Raphson is employed to find the next equilibrium point (x^{j+1}, μ^{j+1}) by solving (2.78) using the approximation from the predictor $(\widehat{x}^{j+1}, \widehat{\mu}^{j+1})$ as the initial point. The acceptable converge error tolerance adopted is 10^{-4} in this thesis, which is suggested by [3].

$$\begin{bmatrix} f(x, \lambda) \\ \varphi(x, \lambda) \end{bmatrix} = 0 \tag{2.78}$$

Step Size Control

Step size control is one of the key factors affecting the efficiency and accuracy of a continuation method. A larger step length suggests a faster operation speed but may lead to convergence problems. On the other hand, a small step size results in a more accurate system equilibrium curve but at the cost of a longer simulation time. This thesis uses the adaptive steps developed from [45], where the step size is determined by (2.79) within the specified limits.

$$o^{j+1} = o^j N_{\text{opt}}/N^j, \; o_{\text{mins}} \leq o^j \leq o_{\text{max}} \tag{2.79}$$

In (2.79), N_{opt} is the optimal number of corrector iterations. As is suggested in [3], N_{opt} is selected as 10 in this thesis, considering an relative error tolerance of 10^{-2}. N^j is the number of iterations needed in the current continuation step. o_{mins} and o_{max} are the minimum and maximum limits of the step. In this thesis, the minimum limit is 10^{-4} as the acceptable relative convergence error used in the corrector step is 10^{-4}. The maximum step limit is set at 10^{-2} since the relative error tolerance adopted in this thesis is 10^{-2}.

2.4.2.3 Error Thresholds in the Continuation Method

The acceptable relative error tolerance used in this thesis is based on the typical relative error tolerances used in the continuation method, which are typically around 10^{-2} [3]. The accuracy of results from the continuation method is mainly affected by two factors: the convergence tolerance set in the corrector step and the step size control. The convergence tolerance determines how well the points generated from the corrector fit the system equilibrium trajectory. In this thesis, this type of relative tolerance is set at around 10^{-4}, enabling the system equilibrium curve to be traced accurately. The step size indicates the distance between the fixed points that form the system equilibrium trajectory. A smaller step size results in a faster computation speed but a larger distance between the two adjacent equilibrium points, which leads to less accurate results. As shown in (2.79), the step size varies from 10^{-4} to 10^{-2}. Combined with the adopted pseudo-arc-length parameterisation technique, the relative difference between equilibrium points from the continuation method may be up to 10^{-2} in magnitude.

2.4.2.4 An Example of the Application of Bifurcation in Power System Stability Analysis

This section offers an example of an application of bifurcation theory in power system stability analysis. Figure 2.13 presents the network diagram of a small test system [49].

The test system consists of three buses: a generator bus (Bus 1), a load bus (Bus 2) and a slack bus (Bus 3). The generator at Bus 1 is modelled using a second-order model, which includes the swing equation of the generator. The generator model is provided as (2.80)–(2.81).

$$\frac{d\delta_m}{dt} = \omega \tag{2.80}$$

$$M\frac{d\omega}{dt} = -D_m\omega + P_m + E_m V Y_m \sin(\delta - \delta_m - \theta_m) + E_m^2 Y_m \sin(\theta_m) \tag{2.81}$$

In (2.80)–(2.81), δ_m is the rotor angle of the SG connected to Bus 1, ω is the SG frequency; P_m is the mechanical power of the SG; E_m is the voltage magnitude at

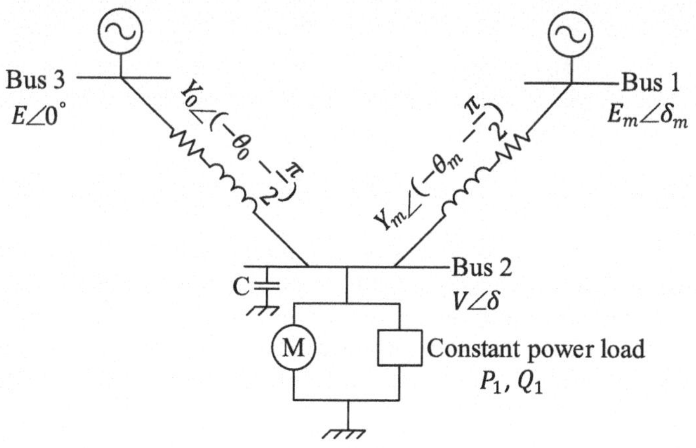

Fig. 2.13 Three-bus test network diagram

Bus 1; $V\angle\delta$ is the voltage at Bus 2; $Y_m\angle(-\theta_m - \frac{\pi}{2})$ is the line admittance between Bus 1 and Bus 2; $Y_0\angle(-\theta_0 - \frac{\pi}{2})$ is the line admittance between Bus 2 and Bus 3; D_m is the damping factor of the SG; M is the SG inertia constant.

In the original system model from [49], a second-order model is used for SG modelling, which is presented in this section. It should be noted that the studies performed in the following chapters will adopt the fourth-order SG model developed in Sect. 2.1.1.2.

The load connected to Bus 2 includes a dynamic induction motor model with a constant PQ load in parallel. The active and reactive power demand of the motor depends on the voltage magnitude and frequency, as shown in (2.82)–(2.83).

$$P = P_0 + P_1 + K_{p\omega}\dot{\delta} + K_{pv}(V + T\dot{V}) \tag{2.82}$$

$$Q = Q_0 + Q_1 + K_{q\omega}\dot{\delta} + K_{qv}V + K_{qv2}V^2 \tag{2.83}$$

In (2.82)–(2.83), P_0, Q_0 are the constant active and reactive power demands of the induction motor; P_1 and Q_1 represent the constant PQ load; $K_{p\omega}$ is the coefficient between the active power and angle; K_{pv} is the coefficient between the active power and voltage magnitude; $K_{q\omega}$ is the coefficient between the reactive power and angle; K_{qv} is the first-order coefficient between the reactive power and voltage magnitude; K_{qv2} is the second-order coefficient between the reactive power and voltage magnitude.

As shown in Fig. 2.13, there is a fixed capacitor C connected to Bus 2 to improve the voltage profile of the system. Instead of including the dynamics of the capacitor in the system model and increasing modelling complexity, it is convenient to account for the capacitor by adjusting E_0, Y_0 and θ_0 based on the Thevenin equivalent circuit. The adjusted values E_0', Y_0' and θ_0' are shown in (2.84)–(2.86).

$$E_0' = E_0/(1 + C^2Y_0^{-2} - 2CY_0^{-1}\cos(\theta_0))^{\frac{1}{2}} \tag{2.84}$$

$$Y_0' = Y_0/(1 + C^2Y_0^{-2} - 2CY_0^{-1}\cos(\theta_0))^{\frac{1}{2}} \tag{2.85}$$

$$\theta_0' = \theta_0 + \tan^{-1}\left\{\frac{CY_0^{-1}\sin(\theta_0)}{1 - CY_0^{-1}\cos(\theta_0)}\right\} \tag{2.86}$$

Hence, the overall power system model can be represented by the four differential equations shown in (2.87)–(2.91) [49, 50]. In general, the system model consists of four state variables corresponding to the generator rotor angle (δ_m), the generator angular speed (ω), the load voltage magnitude (V), and angle (δ), respectively.

$$\frac{\mathrm{d}\delta_m}{\mathrm{d}t} = \omega \tag{2.87}$$

$$M\frac{\mathrm{d}\omega}{\mathrm{d}t} = -D_m\omega + P_m + E_m V Y_m \sin(\delta - \delta_m - \theta_m) + E_m^2 Y_m \sin(\theta_m) \tag{2.88}$$

$$\begin{aligned}K_{q\omega}\frac{\mathrm{d}\delta}{\mathrm{d}t} = & -K_{qv2}V^2 - K_{qv}V + E_0'Y_0'V\cos(\delta + \theta_0') \\ & + E_m Y_m V \cos(\delta - \delta_m + \theta_m) \\ & - (Y_0'\cos(\theta_0') + Y_m\cos(\theta_m))V^2 - Q_0 - Q_1\end{aligned} \tag{2.89}$$

$$\begin{aligned}TK_{q\omega}K_{pv}\frac{\mathrm{d}V}{\mathrm{d}t} = & K_{p\omega}K_{qv2}V^2 + (K_{p\omega}K_{qv} - K_{q\omega}K_{pv})V \\ & + \sqrt{K_{q\omega}^2 + K_{p\omega}^2}\big[-E_0'Y_0'V\cos(\delta + \theta_0' - \eta) \\ & - E_m Y_m V\cos(\delta - \delta_m + \theta_m - \eta) \\ & +(Y_0'\cos(\theta_0' - \eta) + Y_m\cos(\theta_m - \eta))V^2\big] \\ & - K_{q\omega}(P_0 + P_1) + K_{p\omega}(Q_0 + Q_1)\end{aligned} \tag{2.90}$$

$$\eta = \tan^{-1}(K_{q\omega}/K_{p\omega}) \tag{2.91}$$

For this illustration, the reactive power demand of constant *PQ* load (i.e. Q_1), is considered to be the bifurcation parameter.

With the continuation method developed in the previous section, the system equilibrium trajectory can be obtained, which is presented in Fig. 2.14. The horizontal axis is the bifurcation parameter, Q_1, and vertical axis represents the voltage magnitude at Bus 2. As shown in Fig. 2.14, the solid line represents the system's stable fixed points, the dashed line represents the unstable fixed points, and the dash-dotted line represents the unstable equilibrium points due to HB. In general, the lower part of the equilibrium curve consists of unstable fixed points, and the upper part of the trajectory is formed by stable equilibrium points, except for the unstable section due to HB. Starting from the initial point and following the system equilibrium trajectory, the system will first become unstable via HB_1 at $Q_1 = 10.946$ p.u. and transfer back to a stable status after meeting with HB_2 at $Q_1 = 11.407$ p.u. As the bifurcation parameter, Q_1, increases further, the system will encounter an SNB at $Q_1 = 11.411$ p.u. The system equilibrium trajectory vanishes, and no system equilibrium point can be found, if the bifurcation parameter increases beyond the SNB. Figure 2.14 suggests the SNB represents a structural change in the system equilibrium curve. Initially, there are two fixed points in the system: one stable equilibrium point that attracts system operation toward it and another unstable fixed point working as a repeller that drives the system operation away. With the increment of the bifurcation parameter, these two fixed points move closer and ultimately merge and disappear at the SNB. The HBs represent the emergence and disappearance of a periodic orbit, which can be identified via the eigenvalue analysis of the system dynamic state matrix.

The analytical results from the bifurcation theory can be further verified by time-domain simulation. Two system operating points, i.e. A ($Q_1 = 10$ p.u.), B ($Q_1 = 11$

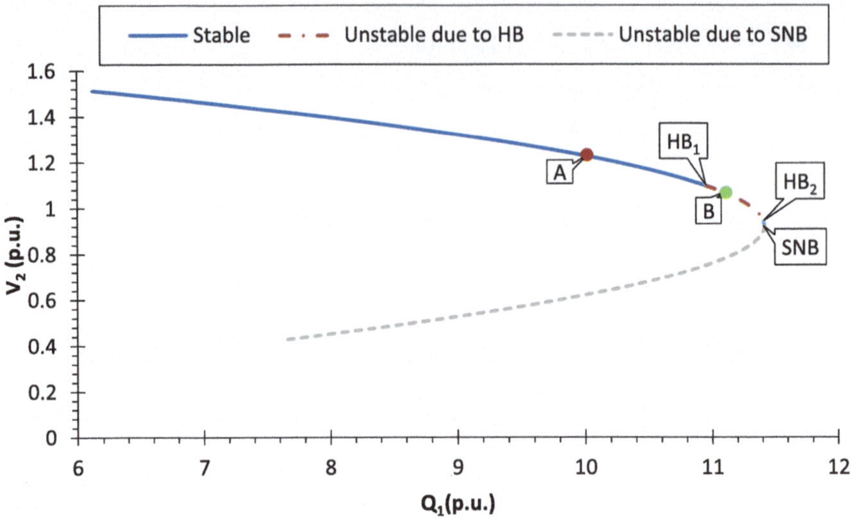

Fig. 2.14 System equilibrium trajectory with Q_1 considered as the bifurcation parameter

p.u.) are selected in Fig. 2.14. Based on the discussion in the previous section, when the system is operating at point A, it is stable. When the operating point moves to point B, the system loses its stability due to HB_1. A small disturbance, in which the active power reference of the SG increases from 1 to 1.1 p.u., is introduced to the system at a time of $t = 10$ s. The frequency of the SG is plotted for the two different operating points. The time-domain simulation results when the system operates at points A and B are shown in Fig. 2.15. For point A, the frequency of the SG stabilises after the changes in the power reference, as is shown in Fig. 2.15a. However, Fig. 2.15b shows that if the system operates at point B, the system becomes unstable after the disturbance. It can be concluded that the time-domain simulation results verify the previous analytical results based on the bifurcation theory. Therefore, this clearly demonstrates that bifurcation theory can be employed to perform the system stability analysis.

2.5 Test Networks

Throughout this thesis, three test networks with modifications have been used: a three-bus system, an IEEE 9-bus system and an IEEE 39-bus system. As the details of the three-bus system have already been presented in Sect. 2.4.2.4, this section will represent the standard AC network for the IEEE 9-bus system and the IEEE 39-bus system. When systems are modified for various studies (e.g., the replacement of an SG with a VSC, or changes in transmission line impedance), they will be discussed in detail on a case-by-case basis to avoid ambiguity.

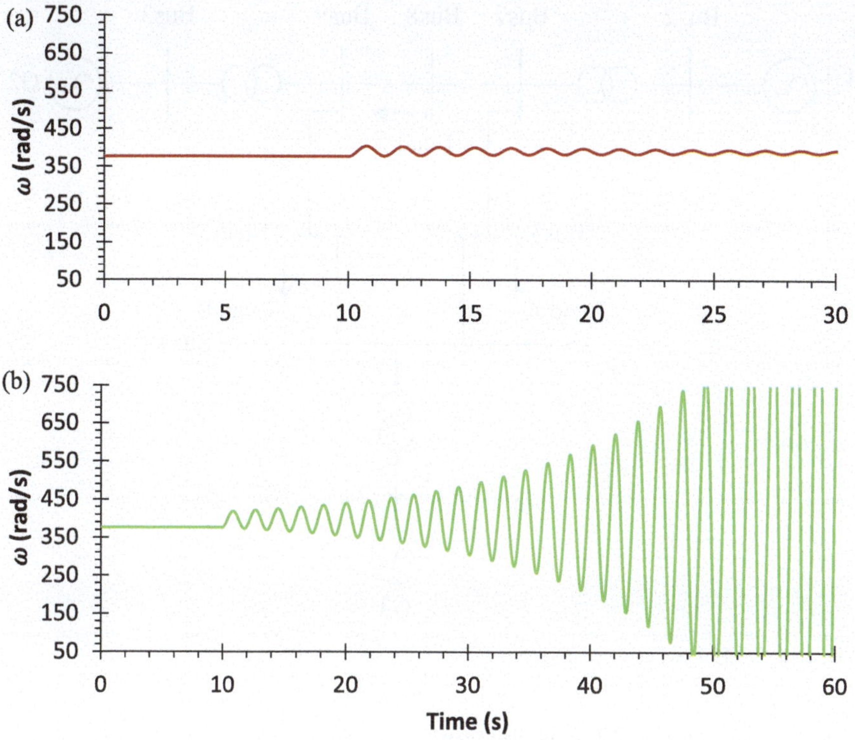

Fig. 2.15 Verification of time domain simulation with **a** point A, and **b** point B

All system data, including the network impedance, the system loading and the generator dynamic data are included in Appendix A. In all cases, the initial power flow for system initialisation will be calculated using the MATPOWER package version 7.0 [8].

2.5.1 The Nine-Bus System

A small, three-machine, nine-bus system is introduced in [44] and has been adopted extensively in previous research for small-disturbance stability analysis [51–53]. Figure 2.16 shows the network diagram of the standard IEEE 9-bus system.

Within the network, G3 is selected as the reference machine that provides the angle reference value for the other generators. All generators are modelled with fourth-order representation combined with the IEEE Type DC1A excitation system.

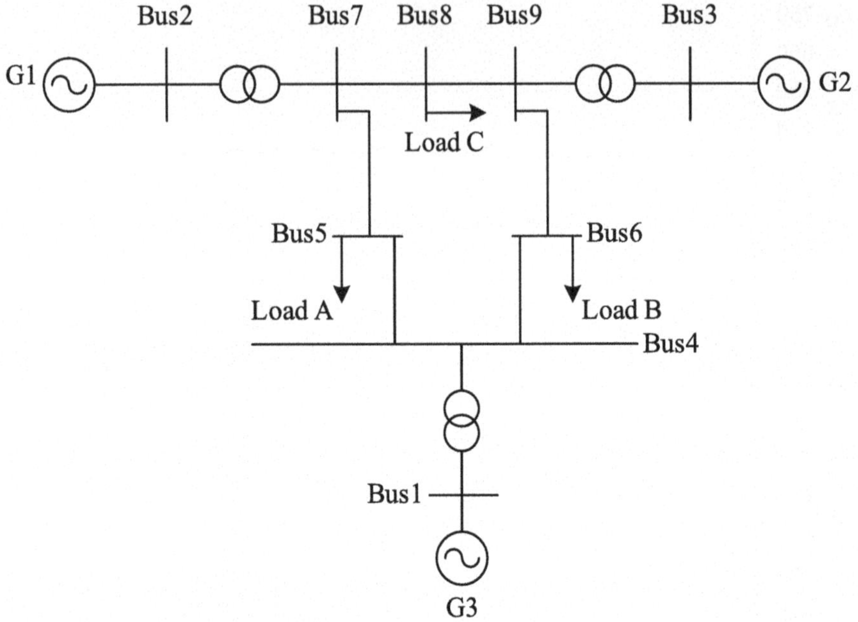

Fig. 2.16 IEEE-9 bus network diagram

2.5.2 *The IEEE 39-Bus System*

A large, ten-machine, thirty-nine-bus network is also used in this thesis for small-disturbance stability analysis. The network diagram is illustrated in Fig. 2.17. The network was developed in [54] and has been extensively used for system small-disturbance stability analysis in past work [55, 56]. The system model represents a reduced-order equivalent model of the New England Test System. G39 is the reference machine that provides the angle reference value. Similar to the IEEE 9-bus system, all generators in the IEEE 39-bus system are represented by the fourth-order model and controlled by the IEEE Type DC1A excitation system.

2.5.3 *Summary of System Models*

Table 2.2 provides a summary of the system models built in this thesis. This includes the names of the system models and the software used to build the models.

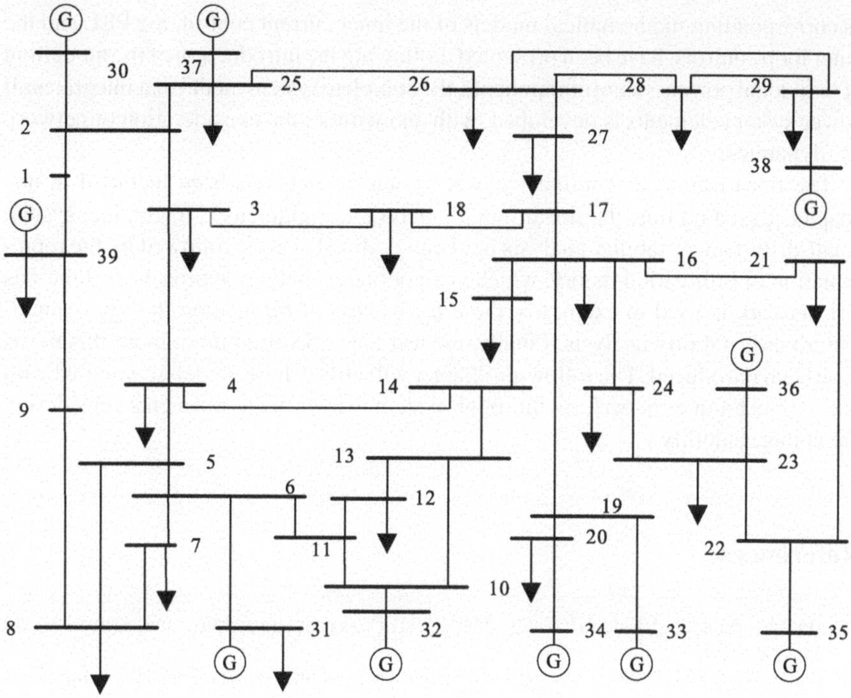

Fig. 2.17 IEEE 39-bus network diagram

Table 2.2 Summary of system models and software used

Model name	Software
Three-bus system (Chaps. 2–6)	MATLAB® (2019a)
Nine-bus system (Chaps. 2–3)	MATLAB® (2019a)
IEEE 39-bus system (Chaps. 2–6)	MATLAB® (2019a), Simulink® (2020a)
Two-VSC test network (Sect. 4.3)	MATLAB® (2019a), Simulink® (2020a)
VSC-HVDC test system (Appendix C)	DIgSILENT (2022)

2.6 Summary

This chapter has presented the fundamental power system modelling and analysis techniques which will be employed to conduct further studies throughout this thesis.

The dynamic characteristics of conventional power system components, alongside their mathematical models, are first presented. These include the modelling details for synchronous generators, their associated controls, branches, loads, and the network (both static and dynamic network models). Following this, a description of the VSC modelling and the cascaded control scheme is presented. The key dynamics, as well as corresponding mathematical models of the inner current control, the PLL and the outer loop controls, have been presented. Following the introduction of the modelling of individual power system components, the complete system model that integrates all power system elements is developed, with and without the consideration of network LC dynamics.

The linearisation of nonlinear power system models has been included in this chapter. Based on this, the linear modal analysis technique used to conduct system small-disturbance stability analysis has been outlined. This is followed by the representation of bifurcation theory, which is a nonlinear analysis approach. A three-bus test network is used to exemplify the effectiveness of bifurcation theory in small-disturbance stability analysis. Finally, the test networks used throughout this thesis have been introduced. The following chapter will utilise these models, combined with analysis techniques to perform thorough system loadability assessments considering the voltage stability.

References

1. (Prabha) Kundur P, Balu NJ, Lauby MG (1994) Power system stability and control, 1st ed. McGraw-Hill
2. Sauer PW, Pai M (2007) Power system dynamics and stability, 1st edn. Stipes Publishing LLC
3. Ajjarapu V (2007) Computational techniques for voltage stability assessment and control, 1st edn. Springer
4. Zhou Y, Ajjarapu V (2005) A fast algorithm for identification and tracing of voltage and oscillatory stability margin boundaries. Proc IEEE 93(5):934–946
5. Venkatasubramanian V, Schättler H, Zaborszky J (1995) Dynamics of large constrained nonlinear systems—a taxonomy theory. Proc IEEE 83(11):1530–1561
6. IEEE Power and Energy Society (2021) IEEE standard definitions for excitation systems for synchronous machines. IEEE Std 421.1-2021 [Revision of IEEE Std 421.1–2007 (Revision of IEEE Std 421.1-1986)]
7. IEEE Power and Energy Society (2016) IEEE recommended practice for excitation system models for power system stability studies. IEEE Std 421.5-2016 (Revision of IEEE Std 421.5–2005), pp 1–207
8. Zimmerman RD, Murillo-s CE (2020) Matpower user's manual Version 7.1
9. Grigsby LL (2017) Power system stability and control, 3rd edn. CRC Press
10. Glover JD, Sarma MS, Overbye TJ (2012) Power system analysis and design, 5th edn. CENGAGE Learning Custom Publishing
11. IEEE Task Force on Load Representation for Dynamic Performance (1993) Load representation for dynamic performance analysis (of power systems). IEEE Trans Power Syst 8(2):472–482
12. Devadason J, Moses P, Fei W (2019) Bifurcation analysis of weak electrical grids considering different load representations. In: 2019 IEEE 7th International Conference on Smart Energy Grid Engineering (SEGE)
13. Mittelstaedt M, Bauer S, Schnettler A (2017) An iterative method for determining the most probable bifurcation in large scale power systems. IEEE Trans Power Syst 32(4):2966–2973
14. Machowski J, Bumby J, Bialek J (2013) Power system dynamics: stability and control, 3rd edn. Wiley
15. Hill DJ (1993) Nonlinear dynamic load models with recovery for voltage stability studies. IEEE Trans Power Syst 8(1):166–176

16. Tuinema BW,Rueda Torres JL, Stefanov AI, Gonzalez-Longatt FM, van der Meijden MAMM (2020) Probabilistic reliability analysis of power systems, 1st edn. Springer International Publishing
17. Vittal V, McCalley JD, Paul AM (2003) Power system control and stability, 3rd edn. Wiley-IEEE Press
18. Karawita C, Annakkage UD (2009) Multi-infeed HVDC interaction studies using small-signal stability assessment. IEEE Trans Power Deliv 24(2):910–918
19. Amin M, Suul JA, D'Arco S, Tedeschi E, Molinas M (2015) Impact of state-space modelling fidelity on the small-signal dynamics of VSC-HVDC systems. In: 11th IET international conference on AC and DC power transmission
20. Randy W et al (2014) Guide for the development of models for HVDC converters in a HVDC Grid. CIGRE Technical Report 604
21. Peralta J, Saad H, Dennetiere S, Mahseredjian J, Nguefeu S (2012) Detailed and averaged models for a 401-level MMC–HVDC system. IEEE Trans Power Deliv 27(3):1501–1508
22. Zhou JZ, Ding H, Fan S, Zhang Y, Gole AM (2014) Impact of short-circuit ratio and phase-locked-loop parameters on the small-signal behavior of a VSC-HVDC converter. IEEE Trans Power Deliv 29(5):2287–2296
23. Shen L, Barnes M, Milanović JV, Bell KRW, Belivanis M (2014) Potential interactions between VSC HVDC and STATCOM. In: Proceedings of 2014 power systems computation conference, PSCC 2014
24. Shah R, Barnes M, Preece R (2017) Characterisation of interaction in an offshore AC grid with large WPPs and VSC-HVDCs. IET Renew Power Gener 11(6):723–731
25. Bayo-Salas A, Van Hertem D, Rimez J, Beerten J (2016) Analysis of control interactions in multi-infeed VSC HVDC connections. IET Gener Transm Distrib 10(6):1336–1344
26. Liu H, Xie X (2019) Comparative studies on the impedance models of VSC-based renewable generators for SSI stability analysis. IEEE Trans Energy Convers 34(3):1442–1453
27. Yazdani A, Iravani R (2010) Voltage-sourced converters in power systems: modeling, control, and applications, 1st edn. Wiley-IEEE Press
28. Wang W (2015) Operation, control and stability analysis of multi-terminal VSC-HVDC systems. Department of Electrical and Electronic Engineering, University of Manchester, Manchester, UK
29. Bjarne Andersen A et al (2005) VSC transmission. CIGRE Technical Report 269
30. Schauder C (1991) Vector analysis and control of advanced static VAr compensators. In: International conference on AC and DC power transmission
31. Skogestad S (2003) Simple analytic rules for model reduction and PID controller tuning. J Process Control 13(4):291–309
32. Jovcic D (2003) Phase locked loop system for FACTS. IEEE Trans Power Syst 18(3):1116–1124
33. Hsieh G-C, Hung JC (1996) Phase-locked loop techniques. A survey. IEEE Trans Ind Electron 43(6):609–615
34. Chaudhuri NR, Majumder R, Chaudhuri B, Pan J (2011) Stability analysis of VSC MTDC grids connected to multimachine AC systems. IEEE Trans Power Deliv 26(4):2774–2784
35. Harnefors L, Bongiorno M, Lundberg S (2007) Input-admittance calculation and shaping for controlled voltage-source converters. IEEE Trans Ind Electron 54(6):3323–3334
36. Wan C, Huang M, Tse CK, Ruan X (2015) Effects of interaction of power converters coupled via power grid: a design-oriented study. IEEE Trans Power Electron 30(7):3589–3600
37. Wang W, Beddard A, Barnes M, Marjanovic O (2014) Analysis of active power control for VSC-HVDC. IEEE Trans Power Deliv 29(4):1978–1988
38. Stamatiou G, Bongiorno M (2016) Stability analysis of two-terminal VSC-HVDC systems using the net-damping criterion. IEEE Trans Power Deliv 31(4):1748–1756
39. Wang X, Blaabjerg F (2019) Harmonic stability in power electronic-based power systems: concept, modeling, and analysis. IEEE Trans Smart Grid 10(3):2858–2870
40. Kuznetsov YA (2004) Elements of applied bifurcation theory, 3rd edn. Springer, New York
41. Venkatasubramanian V, Schättler H, Zaborszky J (1995) Local Bifurcations and feasibility regions in differential-algebraic systems. IEEE Trans Automat Contr 40(12):1992–2013

42. Kwatny HG, Fischl RF, Nwankpa CO (1995) Local bifurcation in power systems: theory, computation, and application. Proc IEEE 83(11):1456–1483
43. Tu Q, Xu Z, Xu L (2011) Reduced Switching-frequency modulation and circulating current suppression for modular multilevel converters. IEEE Trans Power Deliv 26(3):2009–2017
44. Chow JH (1982) Time-scale modeling of dynamic networks with applications to power systems, 1st edn. Springer, Berlin
45. Seydel R (2009) Practical bifurcation and stability analysis, 3rd edn. Springer, New York
46. Venkatasubramanian V, Schattler H, Zaborszky J (1991) A taxonomy of the dynamics of the large power system with emphasis on its voltage stability. Bulk Power Syst Volt Phenom II Volt Stab Secur 9–52
47. Chiang H-D, Flueck AJ, Shah KS, Balu N (1995) CPFLOW: a practical tool for tracing power system steady-state stationary behavior due to load and generation variations. IEEE Trans Power Syst 10(2):623–634
48. Mori H, Yamada S (2002) Continuation power flow with the nonlinear predictor of the Lagrange's polynomial interpolation formula. In: Proceedings of the IEEE power engineering society transmission and distribution conference, pp 1133–1138
49. Dobson I, Chiang HD, Thorp JS, Fekih-Ahmed L (1988) Model of voltage collapse in electric power systems. In: Proceedings of the IEEE conference on decision and control, pp 2104–2109
50. Ajjarapu V, Lee B (1992) Bifurcation theory and its application to nonlinear dynamical phenomena in an electrical power system. IEEE Trans Power Syst 7(1):424–431
51. Revel G, Leon AE, Alonso DM, Moiola JL (2010) Bifurcation analysis on a multimachine power system model. IEEE Trans Circuits Syst I Regul Pap 57(4):937–949
52. Revel G, León AE, Alonso DM, Moiola JL (2008) Bifurcation analysis on a detailed multimachine power system model. In: 40th North American Power Symposium, NAPS2008
53. Liu Y, Liu J, Zhang S, Gareth T, Liu Y (2011) An investigation of limit induced bifurcation in dynamic model of wind power system. In: Asia-Pacific power and energy engineering conference, APPEEC
54. Rogers G (2000) Power system oscillations, 1st edn. Springer, New York
55. Grdenić G, Delimar M, Beerten J (2020) Assessment of AC network modeling impact on small-signal stability of AC systems with VSC HVDC converters. Int J Electr Power Energy Syst 119:105897
56. Fatah Mochamad R, Preece R, Hasan KN (2022) Probabilistic multi-stability operational boundaries in power systems with high penetration of power electronics. Int J Electr Power Energy Syst 135:107382

Chapter 3
Loadability Analysis with Multiple Bifurcations Considered

Voltage stability plays a vital role in modern power system operation, as current systems operate close to their small-disturbance voltage stability limits [1]. This has led to the concept of the system's maximum loadability point, which represents the maximum power loading that a system can handle without violating any stability limits. Any attempt to increase the load beyond the system's maximum loadability point will result in system instability or even system collapse. In general, the system's maximum loadability is recognised as an index that is closely related to voltage stability. Consequently, the prediction, identification, and optimisation of system maximum loadability are important when analysing the voltage stability of a power system.

There has been much interest in the analysis of system loadability using bifurcation theory. As is shown in the literature review in Sect. 1.5.2, early research into system small-disturbance voltage stability and loadability shows that the primary reasons for voltage collapse are related to SNBs and LIBs [2–6]. Other research [7] has demonstrated that HBs can also cause system instability. Because the system always reaches an HB before an SNB, part of the previous research has typically considered the HB as a separate type of bifurcation when discussing system maximum loadability [8, 9]. Although previous studies have used bifurcation theory to systematically investigate a system's maximum loadability points, such research has typically analysed system stability considering each type of bifurcation individually. However, the SNB, LIB, and HB are strongly related to each other. Hence, a thorough investigation of system loadability with the consideration of the SNB, LIB, and HB simultaneously is required.

Compared with the conventional continuation method, optimisation-based techniques offer significant computational savings. A review of the application of optimisation techniques combined with bifurcation theory in analysing the system's

Text and some of the figures in Sects. 3.1–3.3 are adapted from Chen et al. [10], released under a CC BY 4.0 license; https://doi.org/10.1016/j.ijepes.2021.107182

Y. Chen, *Stability Assessment of Power Systems with Multiple Voltage Source Converters*, Springer Theses, https://doi.org/10.1007/978-3-031-63095-8_3

maximum loadability has been presented in Sect. 1.5.2.1. The optimisation-based approaches used in previous research focused on analysing system stability boundaries considering only one bifurcation type, which could be represented directly as a set of mathematical equations. Thus, there is a clear need to develop a new optimisation-based method to investigate system stability boundaries—a method that considers the SNB, LIB, and HB simultaneously. This need was identified in the summary of research gaps in Sect. 1.5.4.

In this chapter, the system loadability concept is updated by simultaneously considering the SNB, LIB, and HB. Five loadability cases, three existing cases from previous work [4, 11, 12] and two new cases based on the updated loadability concept, are proposed and illustrated using a modified three-bus and a nine-bus system. Following this, a renewed system loadability boundary composed of different types of bifurcation points is established considering the SNB, LIB, and HB. The study in the test network demonstrates that with the updated loadability concept, the system loadability boundary is a mix of different types of bifurcations and thus differs from the system boundary considering only one type of bifurcation. This study highlights the importance of considering multiple types of bifurcations simultaneously when analysing the stability region of a system. Otherwise, the results obtained may lead to incorrect estimations of system stability limits.

A new, multi-stage optimisation framework to improve system stability that considers the SNB, LIB, and HB simultaneously is also proposed in this chapter. This is the first novel contribution of this thesis. In general, the developed analysis framework consists of three optimisation loops. The inner optimisation loop determines the system's maximum loadability point and the system loadability boundary, while simultaneously considering different types of bifurcations. The middle loop will identify the shortest distance from any given operating point to the new multi-bifurcation loadability boundary (provided by the inner loop). By optimising the system's shortest loading margin within the parameter space using the outer optimisation loop, the system's overall loadability will be improved. Again, the two small test systems, a three-bus system and a nine-bus system, have been adopted to illustrate the effectiveness of the optimisation-based framework that has been developed. A further brief illustrative example on the larger IEEE 39-bus system is also included to demonstrate the applicability of the method to a more realistic power system.

3.1 Loadability Curves with Multiple Bifurcations

As is discussed previously in Sect. 2.4.2.1, the SNB and LIB represent structural changes in the system equilibrium curve. The HB is a type of bifurcation point related to the emergence and disappearance of a periodic orbit. The occurrence of the HB typically indicates a change in the stability of fixed points, which can be identified using modal analysis. Hence, the SNB and LIB are commonly recognised as static bifurcations, and the HB is classified as a dynamic bifurcation. When assessing the system loadability curve, most existing work considers system instabilities due to static bifurcations only. Unlike previous work, the system loadability analysis presented in this chapter considers both types of bifurcations simultaneously. This section summarises the five different loadability curve cases included in this chapter. The three existing loadability cases from past research [4, 11, 12], which consider only the SNB and LIB simultaneously, are first presented. This is followed by the discussion of two new loadability cases with the simultaneous consideration of the SNB, LIB, and HB.

3.1.1 Previous Work

Previous research into the system maximum loadability has typically considered only the SNB and LIB simultaneously, and hence, the loadability curves can be summarised as three cases [8, 11]. In Fig. 3.1a–c, the plots of the system loadability curves are shown. These are plots of the critical voltage in the system as the loading increases. These cases in Fig. 3.1a–c have been investigated in previous research. Note that the axes in Fig. 3.1a–c are dimensionless, as the figures represent the general shapes of system equilibrium curves found in previous work rather than specific numerical examples.

- **Case 1**: The system's maximum loadability point is determined as an SNB without reaching limits. As shown in Fig. 3.1a, as the system loading $(\mu)^{\lambda}$ increases, the voltage across the generator terminals reduces. At a certain point, further increases in system loading result in the loss of system equilibrium. This critical point is identified as an SNB.
- **Case 2**: The system's maximum loadability point is an SNB after reaching limits. The system experiences a limit-induced dynamic bifurcation (LIDB) before reaching the SNB, as shown in Fig. 3.1b. Theoretically, based on the definition in [13], an LIDB is not a type of bifurcation point. In the LIDB case, the new stable equilibrium is topologically conjugated locally to the previous stable equilibrium, and system stability is not changed before and after the limit. However, previous work [11, 14] refers to 'LIDB' as a power system phenomenon, where the system operation point persists and remains stable after reaching the limit. This chapter will adopt the LIDB terminology for consistency with this previous work.
- **Case 3**: The system's maximum loadability point is a static limit point. The system reaches a limit-induced static bifurcation (LISB), and the LISB is the maximum loadability point, as shown in Fig. 3.1c. If a system encounters an LISB, the system equilibrium trajectory moves along an unstable manifold and immediately becomes unstable [11, 12].

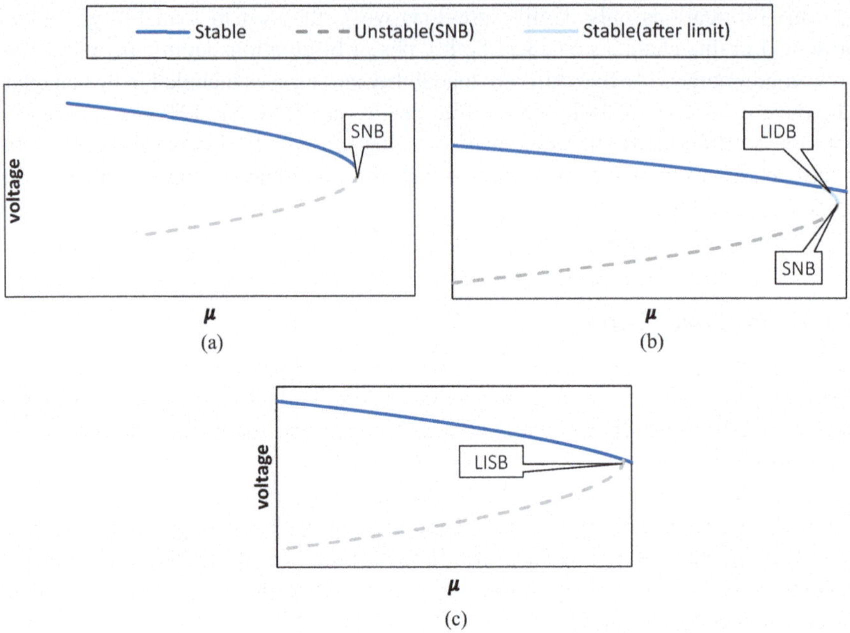

Fig. 3.1 Previous work on the system maximum loadability point with **a** Case 1, **b** Case 2, **c** Case 3, (all axis numbering is removed). Adapted from Chen et al. [10], released under a CC BY 4.0 license; https://doi.org/10.1016/j.ijepes.2021.107182

3.1.2 New Cases Considering SNB, LIB, and HB

Unlike previous analyses, this section investigates system maximum loadability with simultaneous consideration of SNB, LIB, and HB. This results in two new cases, as shown in Fig. 3.2.

- **Case 4**: The system's maximum loadability point is an HB without reaching limits. Following the system equilibrium trajectory, the system may meet with HB points before encountering its SNB, as illustrated in Fig. 3.2a.
- **Case 5**: The system's maximum loadability point is an HB after encountering limits. The system reaches its limit before the original HB is reached. It continues to remain stable on the new equilibrium trajectory (represented by a light blue line) as the bifurcation parameter (μ) increases, as shown in Fig. 3.2b. On this new trajectory, an HB still exists, which becomes the system's maximum loadability point.

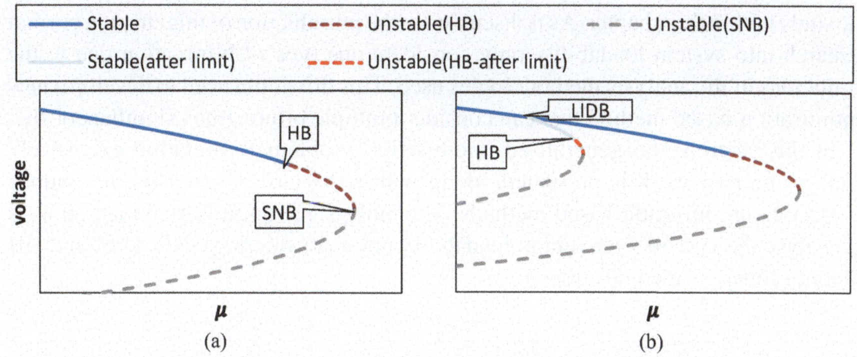

Fig. 3.2 New system loadability cases considering SNB, LIB, and HB simultaneously for **a** Case 4 and **b** Case 5. Adapted from Chen et al. [10], released under a CC BY 4.0 license; https://doi.org/10.1016/j.ijepes.2021.107182

3.1.3 The New System Loadability Boundary

The method used in previous work is based upon one type of bifurcation point only, and thus it assumes the system loadability is smooth and continuous. However, once multiple bifurcation types (SNB, LIB, and HB) are considered, it is possible to have a system loadability boundary which is composed of various portions formed by different bifurcation types. An example of such a composite system loadability boundary will be included in Sect. 3.3.1.5. For these multi-bifurcation loadability boundaries, it is difficult to formulate a single mathematical equation that describes the boundary, as the boundary curve is not guaranteed to be smooth or continuous. This results in a requirement to develop an appropriate optimisation problem formulation that can establish the multi-bifurcation system loadability boundaries.

3.2 A Methodology to Optimise the System Loadability Boundary

The continuation method and the optimisation-based method have been the two commonly used techniques in past work to analyse the system's maximum loadability. In [11], it is shown that both approaches are able to find the system loadability margin accurately. As previously highlighted in the literature review in Chap. 1, optimisation-based methods are more time-efficient compared with the continuation method, but less reliable and informative. The study within this chapter aims to develop a new methodology to improve the system's overall loadability with the simultaneous consideration of multiple bifurcations. The information regarding the system's behaviour towards instability along the system equilibrium curve is of less interest within this study. Hence, the optimisation-based method has been selected for the study within this chapter. As is discussed in the introduction of this chapter, current research into system loadability only considers one type of bifurcation due to the limitations of the analysis methodologies used. This drives the need to develop a new optimisation-based method that can consider multiple bifurcations simultaneously.

In this section, the generalised optimisation problem formulation extensively used in the past work is presented, along with a discussion about the limitations of existing optimisation-based methods. The multi-stage optimisation method used to analyse the system's maximum loadability point considering SNB, LIB, and HB simultaneously is then developed.

3.2.1 Previous Optimisation Problem Formulation

Assuming a power system is represented by DAEs in (2.51), the general form of the optimisation problem formulation to calculate system maximum loadability is expressed by (3.1)–(3.4).

$$\max_{x,y,\mu} \mu \tag{3.1}$$

$$\text{s.t.}\, f(x, y, \mu) = 0 \tag{3.2}$$

$$g(x, y, \mu) = 0 \tag{3.3}$$

$$r_{\min} < r < r_{\max} \tag{3.4}$$

In (3.4), x is the vector of the state variables of the system; y represents the vector of the algebraic variables; f and g are the differential and algebraic equations describe the dynamic characteristics of the system; μ is selected as the bifurcation parameter; and r represents the system parameter constrained by the limits. Constraints (3.2) and (3.3) determine the system equilibrium trajectory (i.e. system loadability curve). Although (3.4) is included in the generalised optimisation problem formulation, it can be used to represent the system's hard operational constraints only. A system's operational constraints are commonly the limits that cannot be violated. Specifically, once the operational constraints are reached, the system operation is terminated, and no solutions to (3.2)–(3.3) exist beyond the limit [12]. An example of an operational constraint in a power system is the voltage operational limit. In the optimisation problem formulation, operational constraints do not directly affect the system matrix, but they do affect system operation. However, in (3.1)–(3.4), there is a lack of information about how the other types of system limits of the system are modelled and included in the optimisation model. These limits include state limits and actuation limits and appear when system state variables or the function of system variables encounter a limit [11, 15]. These limits may impact the system matrix structure and influence system dynamic behaviour. Different from operational limits, the system operating point may still persist after reaching the soft limits. The generation of the excitation field limit is an example of state limits. In summary, the optimisation problem formulation represented by (3.1)–(3.4) models the system's operational constraints, while neglecting the state and actuation limits. Furthermore, the existing generalised optimisation problem formulation focuses on static types of bifurcation only (i.e. SNB and LIB) and fails to consider HB, which is strongly related to oscillatory instabilities in the system.

3.2.2 Proposed Optimisation Problem Formulation (The Inner Optimisation Loop)

To consider the SNB, LIB, and HB simultaneously, a new optimisation problem formulation that combines logic and eigenvalue calculations is proposed, as shown in (3.5)–(3.12). Because the inclusion of the upper (maximum) limit is formed similarly to the lower (minimum) limit, only the maximum limit is presented for brevity.

$$\max_{x,y,\lambda,k,p} \ \mu \tag{3.5}$$

$$\text{s.t. } f(x, y, \mu, k, p) = 0 \tag{3.6}$$

$$g(x, y, \mu, k, p) = 0 \tag{3.7}$$

$$f_1(x, y, \mu, k, p) \cdot f_2(x, y, \mu, k, p) = 0 \tag{3.8}$$

$$f_1(x, y, \mu, k, p) \geq 0 \tag{3.9}$$

$$f_2(x, y, \mu, k, p) \geq 0 \tag{3.10}$$

$$\text{Re}(\text{eig}(\text{Jacobian})) \leq 0 \tag{3.11}$$

$$\kappa(x, y, \mu, k, p) \geq 0 \tag{3.12}$$

In (3.5)–(3.12), p is the controlled parameter, which can be regulated to drive the system to the desired optimisation point; k represents the direction of increase of bifurcation parameters, which will be discussed in the following section; and f and g are the differential and algebraic functions, which are used to represent the system dynamics. The switching functions, f_1 and f_2, are used for modelling the system's state limits; for example, the AVR with over-excitation limiters, which defines the upper limit to the field current [16]. Consider that f_1 and f_2 represent the system dynamics before and after the limit value is reached. Before reaching the limit, f_1 is always equal to zero and hence equates (3.8) to zero. Under normal operation (i.e. before reaching limits), f_2 is greater than zero, which is represented by inequality in (3.10). Once the system increases beyond its limit value, the limiter is activated, and f_2 takes control and forces (3.8) to zero. In this case, f_1 is deactivated, as it has a value greater than zero, which is represented by the inequality in (3.9). A number of system operational constraints are considered. For instance, busbar voltage limits are represented by κ. Once the system meets its operational limits and κ is violated, the system operation will be terminated. eig(Jacobian) in (3.11) is the eigenvalue calculation of the system Jacobian matrix. Equation (3.11) is introduced to ensure all the eigenvalues have negative real parts.

Equation (3.11) is implemented using the constraints defined in (3.13)–(3.17).

$$J^t(x, y, \mu, k, p)\psi_i' - \sigma_i\psi_i' + \omega_i\psi_i'' = 0 \tag{3.13}$$

$$J^t(x, y, \mu, k, p)\psi_i'' - \sigma_i\psi_i'' - \omega_i\psi_i' = 0 \tag{3.14}$$

$$||\psi_i|| - 1 = 0 \tag{3.15}$$

$$\psi'^T \psi'' = 0 \tag{3.16}$$

$$\sigma_i \leq 0 \tag{3.17}$$

for $i = 1 \ldots n$

In (3.13)–(3.17), assuming a system consisting of n unique state variables, $J(x, y, \mu, k, p)$ is the system Jacobian matrix. Assuming that the ith eigenvalue as $\sigma_i + j\omega_i$, $\psi_i = \psi_i' + j\psi_i''$ is the corresponding left eigenvector of the ith eigenvalue. $||\psi_i||$ represents the Euclidean norm of l_i.

In general, (3.5)–(3.10) determine the system loadability point with the consideration of SNB and LIB, which are commonly associated with structural changes in the trajectory of the system equilibrium. Unlike static bifurcation (for example, the SNB and LIB), which is associated with the merger and disappearance of the system equilibrium point, dynamic bifurcation (for example, the HB) is directly determined by system dynamic characteristics. To include Cases 4 and 5 described above, the eigenvalue calculation of the system Jacobian matrix must be included in the optimisation problem formulation. Consequently, (3.13)–(3.17) are included in the proposed optimisation-based method to guarantee that all eigenvalues are located in the stable zone.

In the new optimisation problem formulation, several different types of system limits are considered, and all system loadability scenarios are included. This means that all five cases defined are considered when analysing the system's maximum loadability.

3.2.2.1 The Initialisation of the Proposed Optimisation-Based Method

To use the proposed optimisation-based method to determine the system's maximum loadability, an initial operating point should be calculated for the proposed optimisation problem formulation. Firstly, a stable steady-state point, which is on the system equilibrium trajectory, is selected for power system initialisation. The busbar voltage magnitudes and angles at the selected steady-state point are calculated using the Matpower toolbox on the MATLAB® platform [17]. Based on the initial busbar voltages, the initial values of all system state variables are calculated. The power system is initialised, and the initial values of parameters within (3.5)–(3.12) can be obtained. After the power system initialisation, the initial system Jacobian matrix can be calculated. Then the eigenvalue and corresponding left eigenvector, which is calculated from the initial system Jacobian matrix, can be used for further initialisation of the optimisation problem formulation. That is to say, the parameters within Eqs. (3.13)–(3.17) will be initialised.

3.2.3 Optimisation Framework

As mentioned in Sect. 3.1.3, the renewed system's maximum loadability boundary is composed of several different types of bifurcation points. Hence, it is hard to express the boundary with a simple mathematical equation. Previous optimisation techniques based on equations that directly describe system loadability boundaries are unlikely to be suitable. This chapter proposes a three-loop optimisation framework to incorporate the new system loadability boundary, as shown in Fig. 3.3.

The inner optimisation loop (already discussed in Sect. 3.2.2) will determine the system's maximum loadability point considering the SNB, LIB, and HB simultaneously. The composite loadability boundary consisting of multiple bifurcations can be formed by connecting system loadability points with different directions of increment. Unlike previous studies that were based on the derived mathematical model, the middle optimisation loop in this developed framework is adopted to numerically analyse the obtained system loadability boundary. To improve the overall system loadability, the shortest loading margin from a given operating point is calculated in the middle loop. This distance is then fed into the outer loop and optimised in the control parameter space.

The proposed method contains three optimisation loops, each connected to the others in a nested structure. The inner optimisation loop is the subject of discussion in Sect. 3.2.2, while the details of the outer and middle optimisation loops are explored in the following sections.

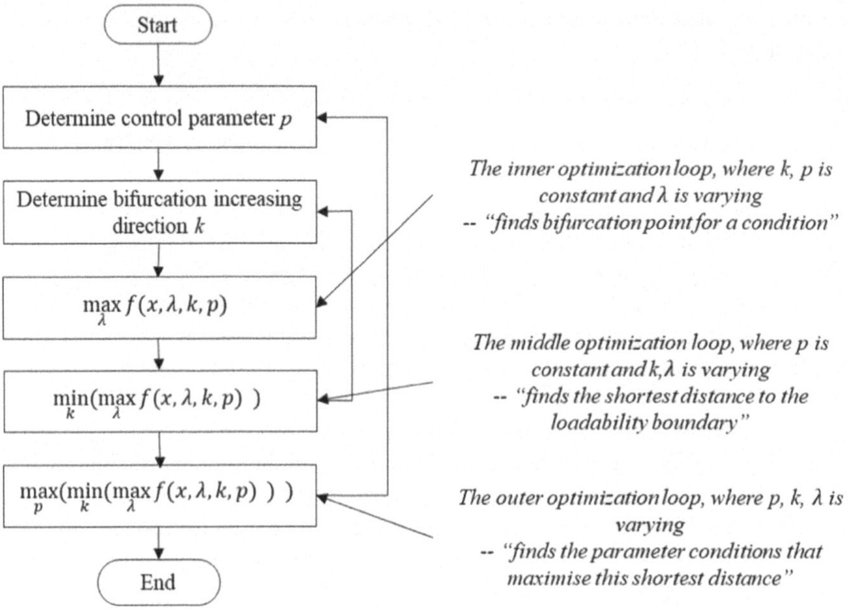

Fig. 3.3 Three-loop optimisation framework. Adapted from Chen et al. [10], released under a CC BY 4.0 license; https://doi.org/10.1016/j.ijepes.2021.107182

3.2.3.1 The Middle Optimisation Loop

The function of the middle loop is to identify the system's shortest loading margin by regulating the direction of increase of the bifurcation parameters. The basic principle behind the function of the middle optimisation loop is shown in Fig. 3.4.

This figure provides an example of a loadability boundary of a power system in which system active (P) and reactive loading (Q) are selected as bifurcation parameters. The set of arrows pointing outwards from the current operating point are the directions of increases of bifurcation parameters. Each direction of increase can be represented by one increasing vector in the two-dimensional plane formed by P and Q. Following the quarter-circle shown in Fig. 3.4, the direction of increase (k) of the bifurcation parameter could change from $k\left(\begin{bmatrix} P \\ Q \end{bmatrix}\right) = \begin{bmatrix} 0 \\ 1 \end{bmatrix} \rightarrow \begin{bmatrix} 1 \\ 0 \end{bmatrix}$. The maximum loadability points are calculated by the inner optimisation loop for every established direction of increase k. In other words, the system will reach the loadability boundary from the current operating point by following each established direction of increase k. The system loadability boundary is created by connecting the obtained maximum loadability points and is represented by the grey line in Fig. 3.4. The Euclidean distance between the current operating point and the point on the loadability boundary can then be measured to determine the loading margin. It is hard to predict the pattern of the bifurcation parameter increase, i.e., power load increase

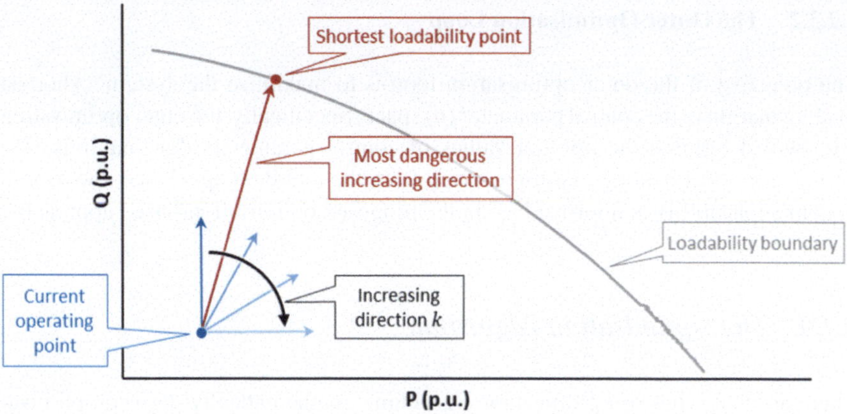

Fig. 3.4 Example of the working principle of the middle loop. Reproduced from Chen et al. [10], released under a CC BY 4.0 license; https://doi.org/10.1016/j.ijepes.2021.107182

in this case [18]. Instead, the middle optimisation loop can determine the shortest loading margin by considering all possible directions of increase. This will identify the direction which, if followed, will intersect the loadability boundary fastest and indicate the worst scenario. Consequently, that direction k and the related system's maximum loadability point are referred to as the most dangerous direction of increase and the system's closest loadability point respectively, which are highlighted in red in Fig. 3.4.

For clarity, only two dimensions, where the system's total active and reactive loading are selected as bifurcation parameters, are shown in Fig. 3.4. A similar analysis process would be applied in cases with higher dimensions, where N bifurcation parameters are considered. Considering a case with N different bifurcation parameters, the middle loop of the proposed methodology will generate a bifurcation parameter direction of increase k in N-dimensional parameter space to determine the system's most critical direction of increase. The two-dimensional illustrative example is generalised to N-dimensions by considering the direction of increase k as an N-dimensional vector rather than as a two-dimensional vector. A detailed application procedure to determine system bifurcations within an N-dimensional bifurcation parameter space will be included in Chap. 4.

In summary, the system's shortest loading margin is established by the middle optimisation loop based on the system's maximum loadability points from the inner loop. The obtained system's shortest loading margin is introduced into the outer loop for further investigation.

3.2.3.2 The Outer Optimisation Loop

The objective of the outer optimisation loop is to maximise the system's shortest loading margin in the control parameter (p) space. Specifically, the outer optimisation loop will determine the optimal values of control parameters that maximise the system's shortest loading margin received from the middle loop. Hence, the overall system's loadability is improved using the proposed optimisation-based approach.

3.2.4 The Optimisation Algorithm

There are two types of optimisation algorithms: mathematically derived optimisation algorithms, such as sequential quadratic programming (SQP) and interior point algorithms, which are used to identify the local minimum value; and evolutionary optimisation algorithms, such as particle swarm optimisation (PSO) and genetic algorithms, which are used to determine the system's global minimum value. The difference between the local and global minimum of the function is shown in Fig. 3.5. A local minimum of the function can be found at point A, where the function value is smaller than that at nearby points. Point B in Fig. 3.5 is a global minimum, where the function value is smaller than all other feasible points.

It should be noted that global optimal values are not always desired. In the proposed optimisation-based analysis framework, the SQP is adopted in the inner optimisation loop to identify the system's local minimum value, while the PSO is used in the middle and outer optimisation loops to compute the system global optimal value. This is discussed more fully in the following sections where the details of SQP and PSO are presented.

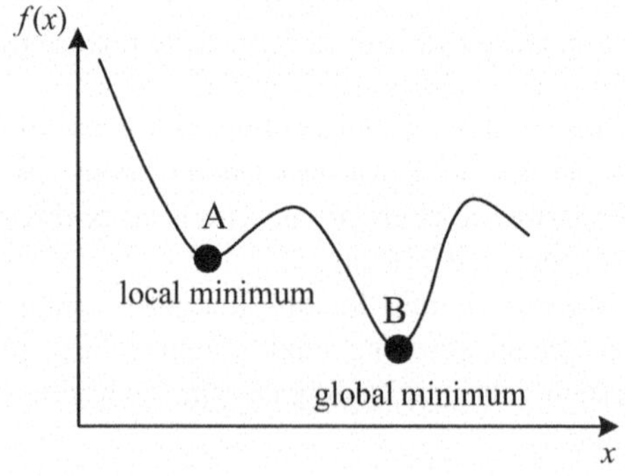

Fig. 3.5 Example of local and global minimum values

3.2.4.1 Sequential Quadratic Programming

SQP has been widely adopted in previous research [19–22] for analysing nonlinear and convex optimisation problems. If the optimisation problem becomes non-convex, then SQP may no longer be suitable, and other algorithms such as genetic algorithms should be adopted. SQP is selected for the inner optimisation loop for its focus on local optimisation. The reason why a local optimisation algorithm rather than a global optimisation algorithm is preferred can be explained by the following scenario.

Consider a system equilibrium curve as shown in Fig. 3.6, where the system contains two HB points before reaching the SNB. Starting from the given initial point and following the system equilibrium trajectory, the system will reach the first HB and become unstable, transferring back to the stable zone after meeting the second HB. Hence, in this case, the first HB is the practical system's maximum loadability point. However, if a global optimisation algorithm is applied, the SNB will be identified as the system's farthest stable equilibrium point and be offered as the optimum result, despite not being acceptable to the system operators. Thus, the local optimisation algorithm SQP was chosen, as the local optimisation result (i.e. the first HB point in this case) is the correct system's maximum loadability point for practical work.

SQP solves the optimisation by updating an approximation of the Hessian of the Lagrangian on each iteration, which is later used to generate a quadratic programming (QP) sub-problem. The solution of the QP sub-problem will determine the direction of the search, which will be further used for a line search procedure [23, 24]. This procedure will be reiterated until the optimal point of the system is found. The general SQP method is as follows, represented by (3.18)–(3.20).

$$\min_{x} f(x) \tag{3.18}$$

$$\text{s.t. } g_i(x) = 0, i = 1 \ldots m_e \tag{3.19}$$

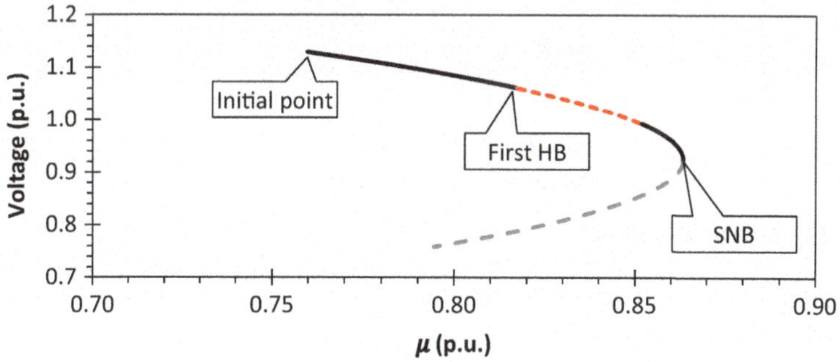

Fig. 3.6 Example of multiple HBs. Adapted from Chen et al. [10], released under a CC BY 4.0 license; https://doi.org/10.1016/j.ijepes.2021.107182

$$g_i(x) \leq 0, i = m_e + 1 \ldots m \tag{3.20}$$

In (3.18)–(3.20), $f(x)$ is the objective function; $g(x)$ is a vector function of constraints; m_e is the number of equality constraints; and m is the total number of constraints.

Given the general problem expressed by (3.18)–(3.20), the formulation of a QP sub-problem can be represented as (3.21)–(3.23) based on a quadratic approximation of the Lagrangian function.

$$\frac{1}{2} d_k^T H_k d_k + \nabla f(x_k)^T d_k \tag{3.21}$$

$$\text{s.t.} \ \left[\nabla g(x_k)\right]^T d_k + g_i(x_k) = 0, i = 1 \ldots m_e \tag{3.22}$$

$$\left[\nabla g(x_k)\right]^T d_k + g_i(x_k) \leq 0, i = m_e + 1 \ldots m \tag{3.23}$$

In (3.21)–(3.23), H_k is the Hessian matrix of the Lagrangian function in (3.24); and d_k is the search direction at iteration k.

$$L(x, \mu) = f(x) + \sum_{i=1}^{m} \lambda_i g_i(x) \tag{3.24}$$

As mentioned above, the implementation of SQP consists of three stages, which are discussed in the following sections.

(1) ***Update the Hessian Matrix of the Lagrangian Function***

At each iteration, a positive definite approximation of the Hessian matrix of the Lagrangian function (B) is updated using the Broyden–Fletcher–Goldfarb–Shanno (BFGS) quasi-Newton method, which is given by (3.25).

$$B_{k+1} = B_k + \frac{q_k q_k^T}{q_k^T s_k} - \frac{(B_k s_k)(B_k s_k)^T}{s_k^T B_k s_k} \tag{3.25}$$

In (3.25), $s_k = x_{k+1} - x_k$, $q_k = \nabla f(x_{k+1}) + \sum_{i=1}^{m} \lambda_i \nabla g_i(x_{k+1}) - \nabla f(x_k) - \sum_{i=1}^{m} \lambda_i \nabla g_i(x_k)$, and λ is the estimated Lagrangian multiplier.

(2) ***Solve the QP Problem***

At each iteration of SQP, a new search direction can be obtained by solving the QP sub-problem defined by (3.21)–(3.23).

(3) ***Determine the Line Search and Merit Function***

The solution of the QP sub-problem determines the new search direction, which is adopted as the basis for the next iteration, as shown in (3.26).

$$x_{k+1} = x_k + \alpha_k d_k \tag{3.26}$$

The step length parameter α_k is determined to produce a sufficient decrease in the augmented Lagrangian merit function. The merit function used in this section is described as (3.27).

$$\Psi(x) = f(x) + \sum_{i=1}^{m_e} r_i g_i(x) + \sum_{i=1}^{m_e} r_i \max[0, g_i(x)] \tag{3.27}$$

In (3.27), r is the penalty parameter, which is set as (3.28).

$$r_i = (r_{k+1})_i = \max_i \left\{ \lambda_i, \frac{(r_k)_i + \lambda_i}{2} \right\} \tag{3.28}$$

It should be noted that the application of SQP in this study is performed using the *fmincon* function in MATLAB® 2019 [25].

3.2.4.2 Particle Swarm Optimisation

PSO is an evolutionary algorithm inspired by the behaviour of a group of organisms. PSO has been widely utilised in power systems to find global optimisation points [26, 27]. In this work, PSO is selected for the middle and outer loops because these optimisation loops require global optimisation results, in contrast with the inner optimisation loop.

The algorithm is performed by considering a group of potential solutions as particles searching the given feasible region in steps. The amassing of particles is called a swarm. Each particle moves within the search region with its own position and velocity. In PSO, each particle will update its velocity at each iteration based on its own experience, the experiences of nearby particles and the global best position among the particles in the swarm to move towards the optimum in the search region [26].

In general, the application of PSO consists of two steps, which are described as follows:

(1) ***Update the Velocity of Particles***

At each iteration of PSO, the velocity of each particle is updated by tracking its best position recorded and the global best position of particles at the current iteration, as shown in (3.29).

$$v_i^{k+1} = \omega v_i^k + c_1 r_1 (p_{\text{best}\,i}^k - X_i^k) + c_2 r_2 (g_{\text{best}}^k - X_i^k) \tag{3.29}$$

In (3.29), i denotes the i th particle of the swarm; k represents the k th iteration; ω is inertial weight, which represents the impact of the previous velocity on the current speed; v_i is the velocity of the particle; x represents the position of the particle; p_{best_i} is the local best position the i th particle has found; g_{best} denotes the global best position of the particle swarm at the current iteration; c_1 and c_2 are the learning factors, which determine the weighting of the speed for an individual particle moving towards its best position and global best position; and r_1 and r_2 are random numbers distributed between [1].

(2) ***Update the Positions of Particles***

Once the velocity of the particles is determined, the positions of the particles can be updated using (3.30).

$$X_i^{k+1} = X_i^k + v_i^{k+1} \tag{3.30}$$

It should be noted that the *particleswarm* function in MATLAB® 2019 is used to perform the PSO in this study.

Based on extensive simulations using the proposed methodology, the developed multi-stage optimisation-based framework demonstrates a high convergence rate, yielding feasible results for most of the tested cases.

3.2.5 Margin Sensitivity

An important step in the proposed framework is to use sensitivity analysis to identify the parameter that is most sensitive to the system loadability boundary. This key parameter will then be used to optimise the system's shortest loading margin in the outer loop. There has been research into the identification of the optimal parameter to improve the system's shortest loading margin [28–31]. However, the sensitivity analysis in the past work focuses on the system loadability boundary formed by only one type of bifurcation. This type of boundary can be directly represented by sets of mathematical equations. Therefore, the sensitivity analysis can be performed directly based on the derivation of the developed mathematical model. However, as was discussed earlier in Sect. 3.1.3, the loadability boundary considered in this chapter consists of multiple bifurcations, and it is extremely difficult to produce a generalised mathematical equation to represent such a boundary. Hence, the previous mathematical, equation-based analysis approach to determining the optimal parameter is no longer suitable here.

Although static bifurcations indicate structural changes in the trajectory of the system's equilibrium, these bifurcation points can be identified through modal analysis, which has been discussed in Sect. 2.4.2.1. One view of loadability is to determine the system's maximum loading point, where at least one of the eigenvalues passes through the imaginary axis with the loading increasing. This eigenvalue is referred to as the critical mode. The parameter that has the most significant impact on the critical eigenvalue will affect the system loadability and further influence the system's shortest loading margin. Consequently, the identification of key parameters that affect the shortest loading margin is approximated as the determination of the parameters that have a significant impact on the critical eigenvalue. Based on the work in [32], eigenvalue sensitivity is defined as (3.31).

$$\frac{\partial \lambda_i}{\partial p} = \psi_i \frac{\partial A_{sys}}{\partial p} \phi_i \tag{3.31}$$

Consider the dynamics of a power system described by a set of ODEs, as in (2.64). The eigenvalue sensitivity can be directly given by (3.31). However, a power system is commonly represented in DAE form, as in (2.51). Although the system state dynamic matrix can be obtained via (2.56), the obtained A_{sys} is a reduced matrix, and hence, it is difficult to directly calculate the term $\frac{\partial A_{sys}}{\partial p}$ in (3.31) for the sensitivity analysis.

This chapter determines the key control parameter used in the outer loop by directly numerically evaluating the parameter derivatives of the real part of the critical eigenvalue with respect to the control parameter. In this sensitivity analysis, one small derivation (denoted as Δ) will be introduced to each control parameter at a time, with the rest of the parameters fixed at their nominal values. The sensitivity measure is then defined as the derivation in the real part of the critical mode ($\Delta\lambda_i$) due to the change in the control parameter (Δp). And (3.32) can be used to express the sensitivity measurement used in this chapter.

$$\frac{\partial \lambda_i}{\partial p} = \frac{\Delta \lambda_i}{\Delta p} \tag{3.32}$$

The parameter that has the largest absolute value from (3.32) will be selected as the key parameter to optimise the system's shortest loading margin.

3.2.6 Application Steps

The application procedure for maximising the system's shortest loading margin in the proposed framework using multiple bifurcations is shown in the flowchart in Fig. 3.7. The detailed application steps are given below:

Step 1: The initial system equilibrium operating point is calculated, and the proposed optimisation problem formulation that was used in the inner loop is initialised as discussed in Sect. 3.2.2.1. The sensitivity analysis based on (3.32) is then performed to determine the key parameter which has the highest sensitivity index value and is used to optimise the system's overall loadability. The initial bifurcation direction of increase and the initial control parameter value are pre-set manually. During the initiation, the pre-set initial value of the direction of increase and control parameter in the inner optimisation loop will be set by the middle and outer optimisation loops, respectively.

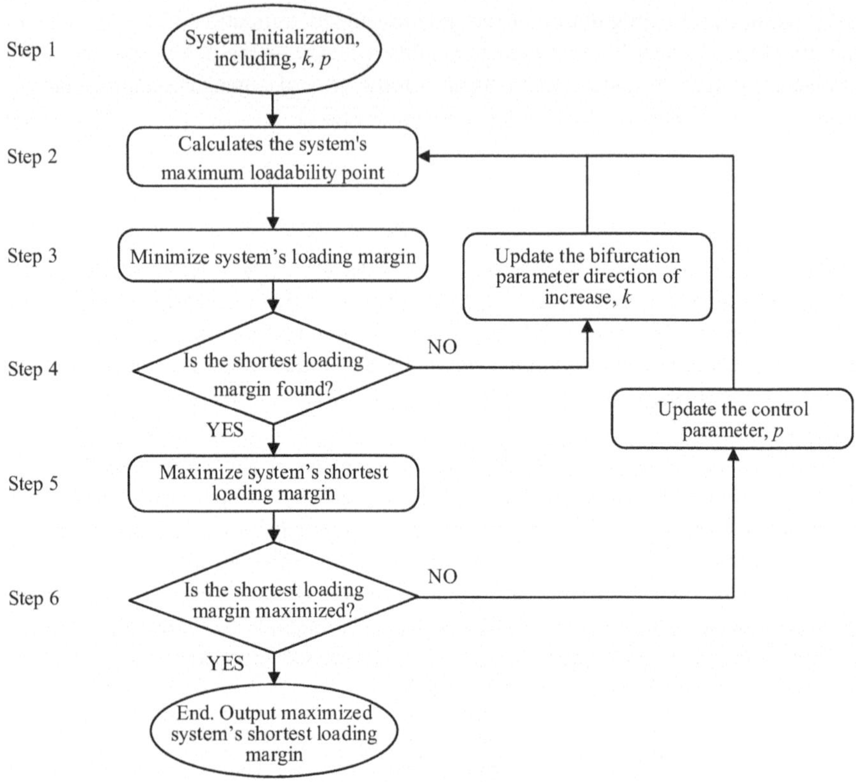

Fig. 3.7 Process flowchart of the proposed methodology. Reproduced from Chen et al. [10], released under a CC BY 4.0 license; https://doi.org/10.1016/j.ijepes.2021.107182

Step 2: The inner optimisation loop calculates the system's maximum loadability point based on the given bifurcation parameter's direction of increase and control parameter setting.

Step 3: The middle optimisation loop calculates and minimises the distance between the current operating point and the results from Step 2 by renewing the bifurcation parameter directions of increase k.

Step 4: The middle optimisation loop continues updating the directions of increase of the bifurcation parameter and creates a loop comprising Step 2 and Step 3 until the system's shortest loading margin is identified.

Step 5: The system's closest loadability point is fed into the outer optimisation loop. The outer optimisation then seeks to maximise the system's shortest loading margin in the control parameter p space.

Step 6: The outer optimisation loop continues updating the control parameter settings and reiterates Step 4 and Step 5 until the system's shortest loading margin is globally maximised in the control parameter space.

The proposed methodology consists of multiple optimisation loops connected in a nested structure, which, in turn, may lead to a long simulation time. Take the simulations conducted using the nine-bus system as an example. In this case, the proposed framework requires approximately 5 h to determine the optimal parameter values for optimising the overall loadability of the system. However, the simulation time is considered acceptable for such an offline analysis.

3.3 Numerical Results

This chapter presents the numerical results from the test systems to illustrate the proposed techniques. The developed analysis framework is first applied to a single-machine, small test system (three-bus system), and then verified in a multi-machine system (nine-bus system). Finally, a large system (IEEE 39-bus system) is used to demonstrate the effectiveness of the proposed analysis framework in a more realistic power system. Within this chapter, the bifurcation parameter (demand loading) is assumed to change in a positive direction (i.e., the load only increases in value). The review shows that both the optimisation-based approach and the continuation method can be employed to identify the bifurcations. Although these two methods have distinct characteristics, advantages, and application objectives, they should produce the same results. Hence, the continuation method is used in this section to verify the results obtained from the developed framework.

3.3.1 The Three-Bus System

The first study into the use of the proposed methodology to optimise the system loadability is completed on the small three-bus system illustrated in Chap. 2. This case study is used to demonstrate the applicability of the developed analysis framework and to verify the necessity of considering multiple types of bifurcations (the SNB, HB, and LIB) simultaneously when assessing the system loadability.

3.3.1.1 Modification of the Test Network

The following modifications are made to the system. Instead of using the second-order generator model from the original paper [33], the fourth-order generator model described in Chap. 2 is used. The IEEE-type DC1A excitation model is adopted to model the generator excitation and AVR behaviours in the test system. The new generator model shares the same power output and capacity as the replaced SG. Within this test network, Bus 3 is modelled as an infinite bus, which represents the external AC network. In general, the network of the three-bus system remains the same as is shown in Chap. 2, except the SG is modelled with a fourth-order model and with an AVR.

3.3.1.2 Application Details

This section details the modelling of the three-bus system and explains the proposed optimisation-based formulation for the power system model. The test network model is represented by sets of ODEs, as shown in (2.87)–(2.91), and fits into the optimisation problem formulation shown in (3.5)–(3.17). f in (3.5)–(3.17) corresponds to a set of ODEs in (2.87)–(2.91). In the three-bus system, f represents the dynamics of the generator, the dynamic load, and so on. Note that the generator excitation behaviour is modelled by f_1 and f_2. As the three-bus system is represented in ODE form, g in this case is a null vector. The equality constraints formed by f build the system's equilibrium trajectory. The system state limit is modelled by the equality and inequality constraints built from the switching function f_1 and f_2, as shown in (3.8)–(3.10). For the three-bus system, the upper voltage limit of the Bus 1 generator field is considered. The f_1 and f_2 functions are used respectively to model the dynamics of the excitation system of the Bus 1 generator before and after the limit value is reached. A series of inequality constraints from function κ are used to model the system's operational limits, such as the voltage and thermal limits. Within this test system, the voltage limits of Bus 1 and Bus 2 are considered.

In (3.5)–(3.17), x represents a vector of the system state variables, representing the dynamic characteristics of each power system component. It should be noted that Bus 3 is considered to be a slack bus, which provides the voltage and frequency reference to the system. Hence, the dynamics of the generator connected to Bus 3 are ignored. Within this study, x is the vector of state variables representing the dynamics of the SG on Bus 1, including the SG and installed AVR dynamics, and the dynamics of inductive motor load on Bus 2, whose state variables are the voltage magnitude and angle on Bus 2. y is a vector of the system algebraic variables. As shown in (2.87)–(2.91), the three-bus system model consists of state variables only, and hence y is a null vector in this case. μ represents the slow-varying bifurcation parameters that drive the system to move from one system equilibrium point to another. In the three-bus system, the bifurcation parameters are the system static active power and reactive power loading on Bus 2. k represents the bifurcation parameter direction of increase, which is explained in Sect. 3.2.3.1. In the three-bus system, k is a two-dimensional vector, reflecting the presence of two bifurcation parameters.p is the controlled parameter used for optimisation to improve system loadability. This could be selected as the power reference setting of the generator, the AVR gain, the reference voltage of the AVR, the reference voltage of the static VAR compensator (SVC), and so on. In this test system, the AVR gain and the reference voltage of the AVR are selected as controlled parameters.

3.3.1.3 Error Thresholds in the Results

The continuation method will be used to verify the results obtained from the proposed methodology. As is discussed in Sect. 2.4.2.3, the acceptable relative error tolerance adopted in the continuation method in this thesis is 10^{-2}. Hence, an acceptable relative error tolerance of 10^{-2} is also used in this chapter.

3.3.1.4 The System's Maximum Loadability Point

In this section, the system's maximum loadability point is found for the case when the bifurcation parameter is the reactive power demand of the constant power load at Bus 2. The various system loadability scenarios discussed in previous sections will be reproduced in the modified three-bus test network with different parameter configurations. The inner optimisation loop from the proposed multi-stage framework is then used to compute the system's maximum loadability point for various scenarios. The results obtained from the inner optimisation loop will be verified by the continuation method.

Figure 3.8 illustrates the loadability curves of the test network, which correspond to Cases 1–4 shown in Sects. 3.1.1 and 3.1.2. Among these four cases, Cases 2 and 3 are produced with the consideration of the maximum excitation voltage limit of the SG connected to Bus 1, $E_{fd\max}^{G1}$. However, Cases 1 and 4 are obtained without consideration of the limit $E_{fd\max}^{G1}$. The horizontal axis in each diagram in Fig. 3.8 is the bifurcation parameter, and the vertical axis represents the generator terminal voltage.

In Fig. 3.8a, where the $E_{fd\max}^{G1}$ is not included, the system directly reaches an SNB at $\mu = 8.633$ p.u. if the generator damping factor (D) increases to 0.5 p.u. This verifies that the loadability scenario shown in Case 1 (discussed in Sect. 3.1.1) exists and can be reproduced using numerical studies.

The loadability curve in Fig. 3.8b is representative of Case 2. In this case, $E_{fd\max}^{G1}$ is considered and set at 0.95 p.u. Figure 3.8b shows that the limit value is reached at $\mu = 8.471$ p.u., and there exists a new system trajectory after the limit, which is represented by the light blue curve. This limit is recognised as an LIDB. As the bifurcation parameter further increases, the system finally reaches the SNB along the new system equilibrium curve at $\mu = 8.489$ p.u. It should be noted that the system equilibrium trajectory after the limit has reduced significantly compared with the original curve. Specifically, the SNB value of the new system loadability curve is smaller than the SNB value located in the original system equilibrium trajectory (i.e. the system will reach the SNB sooner if the limit is met). This is because once the $E_{fd\max}^{G1}$ limit is reached, the AVR can no longer regulate the SG's field voltage and can only produce a constant value of E_{fd}. This means that a number of state variables corresponding to AVR are replaced by algebraic variables with constant values. Under this circumstance, the SG cannot control its terminal voltage via the AVR. Consequently, as loading further increases, the busbar voltage will drop at a

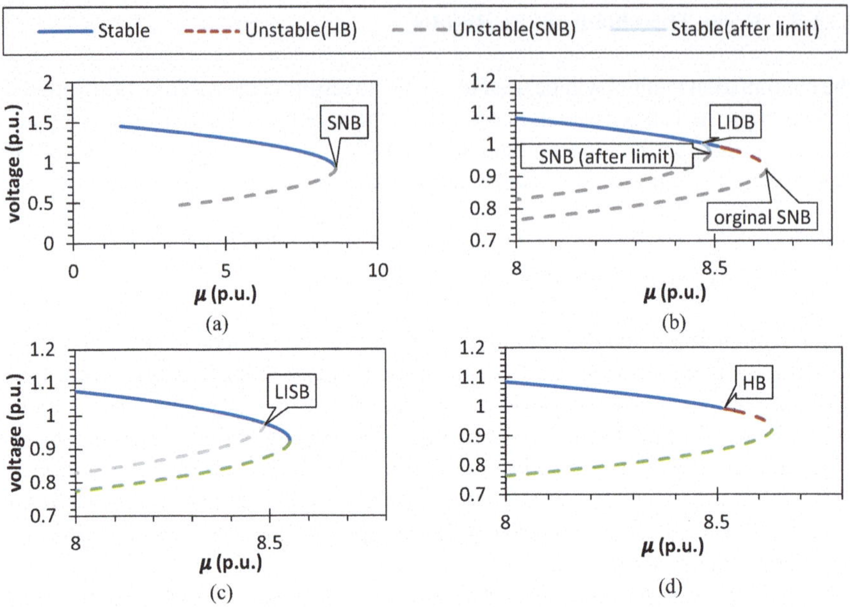

Fig. 3.8 Four cases identified in the three-bus system: **a** Case 1, **b** Case 2, **c** Case 3, and **d** Case 4

faster speed, which subsequently results in a significant reduction in the length of the new system loadability curve, particularly in such a small-sized test network.

The system loadability curve corresponding to Case 3 is reproduced in Fig. 3.8c. In this case, the $E_{fd\max}^{G1}$ set at 0.98 p.u., and the system will encounter an LISB at $\mu = 8.489$ p.u. Hence, the LISB is considered to be the system's maximum loadability point, beyond which no further system fixed point can be established.

Figure 3.8d illustrates the nominal loadability curve of the test network without considering $E_{fd\max}^{G1}$, in which an HB point occurs along the system equilibrium trajectory at $\mu = 8.517$ p.u. In this scenario, HB becomes the system's maximum loadability point. This verifies the loadability scenario in Case 4 (discussed in Sect. 3.1.2).

Case 5 describes a system experiencing an LIDB before meeting with an HB. This is more likely to occur in a multi-machine system. Therefore, the reproduction of Case 5 will be performed in the nine-bus system.

Table 3.1 compares the results of the solutions obtained from the proposed optimisation framework for scenarios matching Cases 1–4. The results demonstrate that the proposed optimisation approach is accurate to within the allowable tolerance (10^{-2}) when compared to the continuation method.

Table 3.1 Comparison of loadability using the continuation and proposed optimisation-based methods for cases 1–4. Adapted from Chen et al. [10], released under a CC BY 4.0 license; https://doi.org/10.1016/j.ijepes.2021.107182

Scenario	Continuation method (p.u.)	Proposed optimisation-based method (p.u.)
1	8.633(SNB)	8.629(SNB)
2	8.471(LIDB), 8.491(SNB)	8.473(LIDB), 8.490(SNB)
3	8.489(LISB)	8.490(LISB)
4	8.517(HB)	8.519(HB)

3.3.1.5 The System Loadability Boundary

In this section, both the active and reactive power of constant power load are considered to be bifurcation parameters. Consequently, the loadability boundary can be obtained by connecting the different system's maximum loadability points. The various loadability curves are first computed to show the necessity of considering different bifurcations simultaneously during the system loadability assessment. Following this, the most dangerous direction of increase is determined using the middle optimisation loop from the proposed framework. The results will be verified using the continuation method. It should be noted that the middle loop contains the inner optimisation loop. Subsequently, the middle optimisation loop with its nested optimisation loop, i.e., the inner optimisation loop, is activated in this section to determine the locally closest loadability point based on obtained the loadability boundary.

Figure 3.9 illustrates the three different system loadability boundaries for SNBs only, SNBs and HBs concurrently, and SNBs, HBs, and LIBs simultaneously. Those three boundary trajectories included in Fig. 3.9 are represented as a red line, a green line, and a blue line, respectively, and stable zones are located below those boundary curves. Figure 3.9 clearly illustrates that the loadability boundary consisting of a single type of bifurcation is different from the boundary which considers multiple bifurcations. This is because the boundary with the simultaneous consideration of different types of bifurcation simultaneously consists of various types of bifurcation. Take the loadability boundary considering the SNB and HB (represented by the red line) as an example. Initially at $k\left(\begin{bmatrix} P \\ Q \end{bmatrix}\right) = \begin{bmatrix} 0 \\ 1 \end{bmatrix}$, the system's maximum loadability points are HB, and hence, the red curve is different from the green line which considers SNBs only. As the bifurcation direction increases, $k\left(\begin{bmatrix} P \\ Q \end{bmatrix}\right) = \begin{bmatrix} 0 \\ 1 \end{bmatrix} \rightarrow \begin{bmatrix} 1 \\ 0 \end{bmatrix}$, the system's maximum loadability point changes from HBs to SNBs, and hence, the previous two different boundaries coincide (i.e., red and green lines merge). Past research has shown that the HB always occurs before the SNB in the system equilibrium curve. Consequently, the system's stable region is reduced when the loadability boundary considering multiple bifurcations is used. This can be further demonstrated by the system loadability boundary when $E_{fd\max}^{G1}$ is considered (set to 0.95 p.u.). This loadability boundary, shown by a blue line in Fig. 3.9, is also a mix of different types of bifurcation. As the bifurcation direction increases $k\left(\begin{bmatrix} P \\ Q \end{bmatrix}\right) = \begin{bmatrix} 0 \\ 1 \end{bmatrix} \rightarrow \begin{bmatrix} 1 \\ 0 \end{bmatrix}$, the system's maximum loadability points change from SNBs to LISBs. With the introduction of $E_{fd\max}^{G1}$, the loadability boundary is further shifted to the left, and the stable region of the system is further reduced. The results presented in Fig. 3.9 clearly demonstrate the necessity to consider multiple types of bifurcation simultaneously when assessing system loadability. Otherwise, the results obtained may lead to a significant loadability assessment error (i.e., an overly optimistic conclusion). Furthermore, the results also verify the discussion in Sect. 3.1.3 suggesting that the renewed system loadability boundary considering the SNB, LIB, and HB is formed from a mix of different types of bifurcation points. Therefore, it is unlikely that this boundary can be expressed by one mathematical equation.

Table 3.2 summarises the normal vector to the hypersurface of the load power, i.e. the most dangerous direction of increase, identified using the two analysis approaches when $E_{fd\max}$ is and is not considered.

It is evident from Table 3.2 that the middle optimisation loop of the proposed analysis framework can be used to determine the system's normal vectors, which are

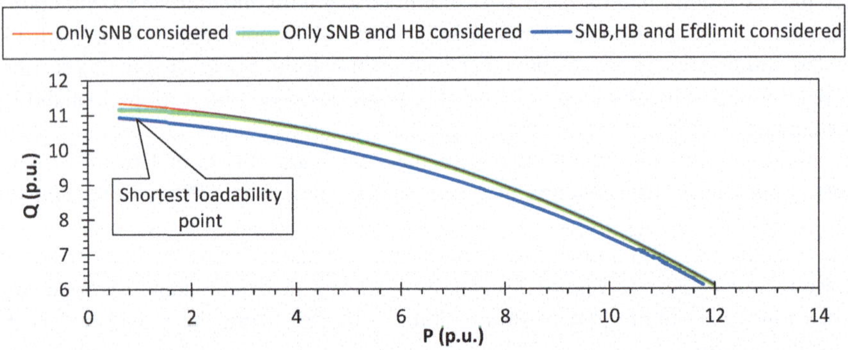

Fig. 3.9 System loadability boundary with different bifurcations considered. Reproduced from Chen et al. [10], released under a CC BY 4.0 license; https://doi.org/10.1016/j.ijepes.2021.107182

Table 3.2 Comparison of the most dangerous direction of increase identified from the continuation and proposed optimisation-based methods

Types of bifurcations considered	Continuation method	Proposed optimisation-based method
SNB, HB	$\begin{bmatrix} 0.000 \\ 1.000 \end{bmatrix}$	$\begin{bmatrix} 0.010 \\ 0.998 \end{bmatrix}$
SNB, HB, and LIB	$\begin{bmatrix} 0.010 \\ 0.999 \end{bmatrix}$	$\begin{bmatrix} 0.010 \\ 0.996 \end{bmatrix}$

similar to those obtained by the continuation method allowing for acceptable relative error (10^{-2}).

3.3.1.6 Optimising the System Loadability Boundary

This section will implement the outer loop to improve the system's overall loadability, i.e. maximise the shortest loading margin in the given parameter space. Sensitivity analysis is conducted to select the most influential parameter to improve the loadability margins. Within the given parameter space, the system's shortest loading margin is optimised using the outer optimisation loop. The outer optimisation loop optimises the shortest loading margin based on the locally closest loadability point computed from the middle optimisation loop, which further requires results from the inner loop. Therefore, all three optimisation loops are used in this section to enhance the system's overall loadability. The continuation method is subsequently employed to compute the shortest loading margin trajectory with respect to the change in the selected parameter in order to test the optimisation results by numerically identifying the optimal parameter setting. The results from the continuation method will be compared with those from the developed outer optimisation loop.

The sensitivity analysis based on (3.15) is performed to identify the most influential parameter affecting the system's shortest loading margin. A comparison of the sensitivity indices of controllable parameters, including generator AVR gain and the reference voltage, is shown in Table 3.3. The table indicates that the reference voltage has a larger sensitivity index than that of the AVR gain. Therefore, the reference voltage is selected as the control parameter. A maximum voltage busbar limit (set to 1.1 p.u.) is considered. The trajectory of the system's closest maximum loadability point with respect to control parameter p is shown in Fig. 3.10. The case shown in the last section, where both SNB and HB are considered, is adopted in this section. In the diagram, the horizontal axis is the generator reference voltage, which is the optimised parameter. The vertical axis represents the system's minimum load margin. The generator reference voltage varies from 1.000 p.u. to 1.150 p.u.

As shown in Fig. 3.10, the system's minimum load margin increases from 5.025 p.u. to 5.523 p.u. (a 9.9% improvement) as the reference voltage varies from 1.000 p.u. to 1.104 p.u. A higher reference voltage setting will always result in a higher busbar voltage at the generator terminal. In this small test system, with a high busbar voltage, the system loadability point will be extended, and the distance to the system's locally closest loadability point will be improved. However, due to the busbar voltage limits, there is a limit to how much the closest boundary point can

Table 3.3 Sensitivities of loading margin to parameters for the three-bus system. Reproduced from Chen et al. [10], released under a CC BY 4.0 license; https://doi.org/10.1016/j.ijepes.2021.107182

Generator number	Reference voltage	AVR gain
1	0.8354	1.342×10^{-4}

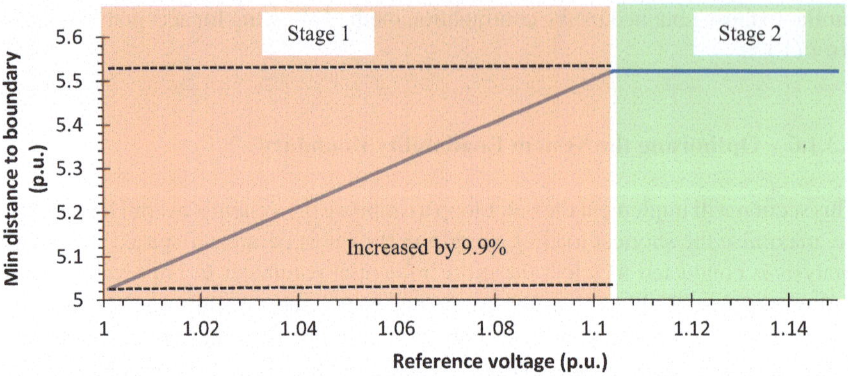

Fig. 3.10 Closest system loadability point with a varying reference voltage for the three-bus system. Adapted from Chen et al. [10], released under a CC BY 4.0 license; https://doi.org/10.1016/j.ijepes.2021.107182

be improved. That is why the minimum distance to the boundary becomes constant after the generator reference voltage reaches 1.104 p.u. Ultimately, the optimal generator reference voltage obtained from the proposed optimisation-based framework is 1.104 p.u. This agrees with the results obtained using the continuation method.

The system loadability Cases 1–4, discussed in previous sections, have been reproduced in the modified three-bus test system. It has been demonstrated in Sect. 3.3.1.5 that when the SNB, HB, and LIB are considered simultaneously, the system loadability boundary is a composite curve consisting of different types of bifurcations. This results in a more complex case to investigate than when only a single bifurcation type is considered. It has been shown that system loadability assessments considering only a single type of bifurcation may result in misleading or incorrect conclusions. Additionally, the studies have clearly shown the effectiveness of the proposed multi-stage optimisation-based methodology. With the application of the inner, middle, and outer loops, the proposed methodology has the capability of identifying the system's maximum loadability point, determining the locally closest loadability point, and optimising the shortest loading margin respectively.

3.3.2 *The Nine-Bus System*

The results from the small three-bus test network presented in Sect. 3.3.1 provide a general picture of the loadability curves and the boundary of the system with multiple bifurcations. It also demonstrates the effectiveness of the proposed optimisation-based framework for improving the system loadability. In this section, various system loadability curves based on the updated loadability concept are further investigated in the nine-bus system, which consists of multiple machines. The proposed optimisation-based methodology is also adopted in the nine-bus system to further demonstrate its suitability for use in a multi-machine system.

3.3.2.1 Application Details

In the nine-bus system, the generator connected to Bus 3 is selected as the reference machine. The nine-bus system can be represented by the DAE formulation shown in (2.51) and fitted into the proposed optimisation problem formulation represented by (3.5)–(3.17). In (3.5)–(3.17), f represents the differential equations of SGs. The generator field voltage upper limits of generators are considered and modelled by f_1 and f_2, which represent the dynamics of the excitation system of each generator before and after reaching the limits. It should be noted that as this section focuses on the small-disturbance voltage stability of power systems, the LC network dynamics have been neglected. Hence, a static network model is adopted in the nine-bus system model, and function g in (3.5)–(3.17) represents the network power flow equations. Like the three-bus system, the voltage limit at each busbar is represented by κ in the nine-bus system.

In terms of the details of the symbols, x is a vector that includes the state variables of generators and their associated AVRs; y is a vector of system algebraic variables, which, in this test system, are the voltage magnitude and angle at each busbar. The bifurcation parameters, μ, are the system's total active and reactive power loading in the nine-bus system. In this test network, the direction of increase in k is a two-dimensional vector, reflecting the presence of two bifurcation parameters. The AVR gains of the generators and the reference voltages of the AVRs are selected as the controlled parameter p in the nine-bus system.

3.3.2.2 The System Maximum Loadability Point

In the nine-bus system, it is typical to consider the total system loading as a bifurcation parameter. Case 4 and Case 5 discussed previously can be reproduced and illustrated by the test system, as shown in Fig. 3.11. The horizontal axis in each diagram is the reactive power load, and the vertical axis represents the voltage at Bus 2. This section follows a similar analysis process to Sect. 3.3.1.4, in which the inner loop of the proposed methodology is used to compute the maximum loadability point in the nine-bus system and is compared with the results using the continuation method.

It is clear from the system loadability curve in Fig. 3.11a that when $E_{fd\max}$ is not considered in the generator, the system reaches an HB at μ=3.831 p.u. This verifies Case 4 as discussed in Sect. 3.1.2. If a limit of $E_{fd\max}$= 3 p.u. is implemented, the system still encounters an HB, but at a new point of $\mu = 3.386$ on the new (limited) system trajectory after meeting with $E_{fd\max}$, as shown in Fig. 3.11b. The results in Fig. 3.11b correspond to Case 5, where the system equilibrium curve represented by the light blue line still exists after $E_{fd\max}$ is encountered. Unlike in Case 2, the HB still exists in the new equilibrium solution curve. Hence, the new HB in this case

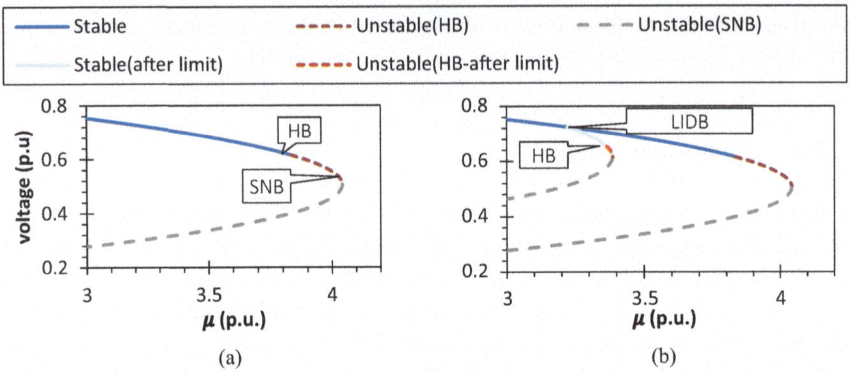

Fig. 3.11 Two cases identified in the nine-bus system: **a** Case 4 and **b** Case 5. Adapted from Chen et al. [10], released under a CC BY 4.0 license; https://doi.org/10.1016/j.ijepes.2021.107182

becomes the system's maximum loadability point. This is because in multi-machine systems if only one of the generators' excitation voltages reaches its limit, a large number of state variables remain in the model. It is thus likely that the HB will remain in the new system equilibrium curve. However, if multiple AVRs cannot control the terminal voltages, and a large proportion of the system's differential equations vanish in the model, the HB point may disappear from the system equilibrium curve.

The results (i.e. maximum loadability points) of the system loadability curves corresponding to Case 4 and Case 5 are illustrated in Table 3.4. The table clearly demonstrates that both methodologies produce similar results within an acceptable tolerance (10^{-2}).

Apart from the system's maximum loadability represented by a single type of loadability case, the loadability trajectory of a power system can be a mix of various types of loadability cases. For example, a multi-machine system may experience several LIDBs before subsequently encountering a system operational limit, as is illustrated in Fig. 3.12. To generate the results shown in Fig. 3.12, the $E_{fd\,\max}$ values of the two generators are set at $E_{fd\,\max}^{G1} = 1.8$ p.u., $E_{fd\,\max}^{G2} = 1.6$ p.u. for (a), and at $E_{fd\,\max}^{G1} = 3$ p.u., $E_{fd\,\max}^{G2} = 3$ p.u. for (b).

For the first case (a), the system undergoes an LIDB due to the $E_{fd\max}$ limit and reaches a further LIDB. Finally, the system reaches an SNB. If the $E_{fd\max}$ limits are increased to 3 p.u. at each generator for case (b), the system's maximum loadability point becomes an LISB instead, preceded by an LIDB, as shown in Fig. 1.12b. It is

Table 3.4 Comparison of loadability using the continuation and proposed optimisation-based methods for Case 4 and Case 5. Adapted from Chen et al. [10], released under a CC BY 4.0 license; https://doi.org/10.1016/j.ijepes.2021.107182

Scenario	Continuation method (p.u.)	Proposed optimisation-based method (p.u.)
4	3.830 (HB)	3.830 (HB)
5	3.386 (HB)	3.386 (HB)

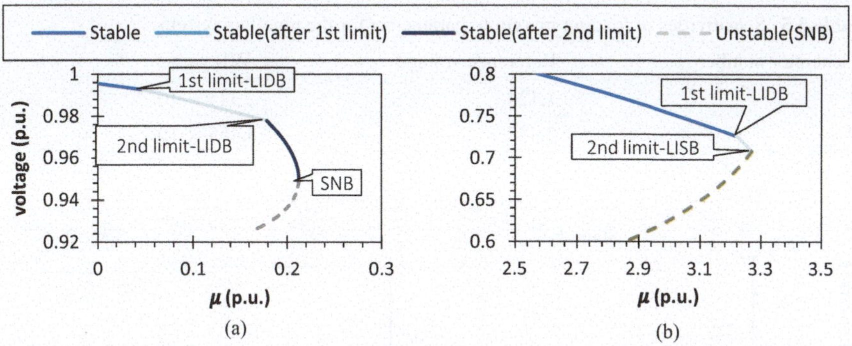

Fig. 3.12 Cases where the $E_{fd\,\max}$ of both generators is considered, with **a** $E^{G1}_{fd\,\max} = 1.8$ p.u., $E^{G2}_{fd\,\max} = 1.6$ p.u., and **b** $E^{G1}_{fd\,\max} = 3$ p.u., $E^{G2}_{fd\,\max} = 3$ p.u.. Adapted from Chen et al. [10], released under a CC BY 4.0 license; https://doi.org/10.1016/j.ijepes.2021.107182

evident that with a higher value of $E_{fd\max}$ limit, the system is more likely to experience the LISB rather than the LIDB.

It should be noted that, as only two generators are considered in the nine-bus system, the system can only experience up to two LIDBs before meeting the system loadability limit. If a large-sized power system with a number of SGs is considered, the system could experience multiple LIDBs before encountering the operational limit.

3.3.2.3 Optimising the System Loadability Boundary

Similar to the application in the three-bus test system, the outer loop of the developed methodology is used in this section to improve the system's loadability boundary within the given parameter space. The sensitivity indices of a small number of different control parameters are listed in Table 3.5. Based on the results, the generator reference voltage of the second generator, which has the largest sensitivity index value, is chosen as the control parameter to optimise the system's closest loadability point. The trajectory of the system's closest loadability point with changing reference voltage is illustrated in Fig. 3.13. The plot in Fig. 3.13 indicates there are two stages in the result. In the first stage, the system minimum load margin increases from 2.289 p.u. to 2.482 p.u. as the generator reference voltage increases from 1.000 p.u to 1.116 p.u. In the second stage, the system's minimum load margin flattens out at 2.482 p.u., where the reference voltage exceeds 1.116 p.u.

In this case, the system's shortest loadability point is always an HB. In the first stage, the HB is caused by the AVR of the first generator based on the participation factor analysis. During this stage, a higher reference voltage will lead to a higher busbar voltage, which, in turn, will delay the occurrence of the HB. By increasing the reference voltage during the first stage, the system's shortest loading margin is improved by 8.4%. During the second stage, no further changes occur once the busbar

Table 3.5 Sensitivities of loading margin to parameters for the nine-bus system

Generator number	Reference voltage	AVR gain
1	1.124	2.345×10^{-4}
2	2.457	5.465×10^{-5}

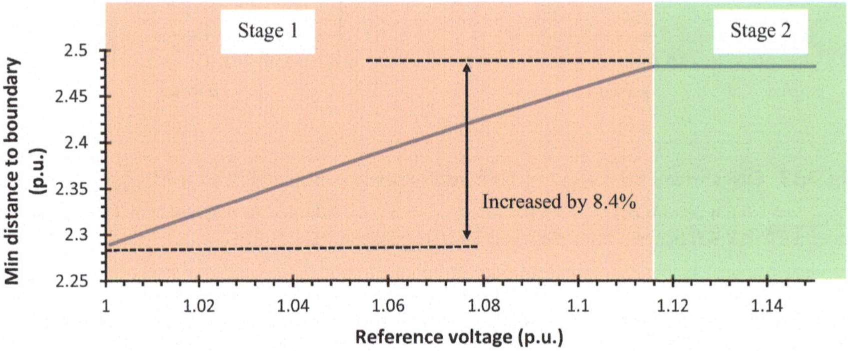

Fig. 3.13 Closest system loadability point with a varying reference voltage for the nine-bus system

voltage limit is encountered, which is why the system's minimum loading margin is constant during the second stage. Therefore, the reference voltage of 1.116 p.u. is the optimal value achievable without violating busbar voltage limits. The value from the proposed optimisation-based framework is 1.116 p.u., which is the same as this obtained from the continuation method.

Various system loadability curves have been computed in the nine-bus system. It is evident from the results that a multi-machine power system may experience several LIDBs before meeting the final maximum loadability point. This confirms the discussion in the introduction: there are interactions between the SNB, HB, and LIB. Additionally, the study within this section has demonstrated the suitability of the developed methodology in a multi-machine system.

3.3.3 The IEEE 39-Bus System

Having established the applicability of the proposed analysis framework on the 3-bus and 9-bus test systems, this methodology is tested on the larger IEEE 39-bus test network. This allows for a comprehensive assessment of its performance on a test network which is closer to a real power system.

3.3.3.1 Application Details

A similar application process to that used for the nine-bus system (described in Sect. 3.3.2.1) is adopted to fit the mathematical model of the IEEE 39-bus system into the proposed optimisation problem formulation.

3.3.3.2 Optimising the System Loadability Boundary

Similar to the previous two test systems, the total system loading demand is considered as the bifurcation parameter within the IEEE 39-bus system. Within this section, the outer loop is used to optimise the system's shortest loading margin in the IEEE 39-bus system. Table 3.6 illustrates the sensitivity indices of a number of considered control parameters (i.e., the reference voltage and AVR gain) of different generators. The sensitivity results indicate that the G39 generator reference voltage has the largest sensitivity index value and, hence, is selected as the control parameter to optimise the system's shortest loading margin. The trajectory of the system's shortest loading margin, as the G39 reference voltage increases, is shown in Fig. 3.14. As illustrated by Fig. 3.14, the trajectory can be divided into two main stages. In the first stage, the system's minimum load margin increases from 3.227 p.u. to 3.400 p.u. as the G39 reference voltage increases from 1.000 p.u to 1.111 p.u. In the second stage, the system's minimum loading margin is constant with further increases in the reference voltage above 1.111 p.u.

During the first stage, the system's minimum loading margin increases, as a higher reference voltage will result in a higher busbar voltage, which will further improve system loadability. Hence, the system's minimum loading margin is improved by 5.4% in the first stage. However, in the second stage, the system's minimum loading

Table 3.6 Sensitivities of loading margin to parameters for the IEEE 39-bus system Reproduced from Chen et al. [10], released under a CC BY 4.0 license; https://doi.org/10.1016/j.ijepes.2021.107182

Generator number	Reference voltage	AVR gain
30	0.164	1.257×10^{-4}
31	0.112	1.171×10^{-4}
32	0.078	2.587×10^{-4}
33	0.015	3.575×10^{-4}
34	− 0.024	1.687×10^{-4}
35	− 0.018	1.894×10^{-4}
36	− 0.006	2.564×10^{-4}
37	− 0.014	1.153×10^{-4}
38	0.021	2.574×10^{-4}
39	0.257	3.561×10^{-4}

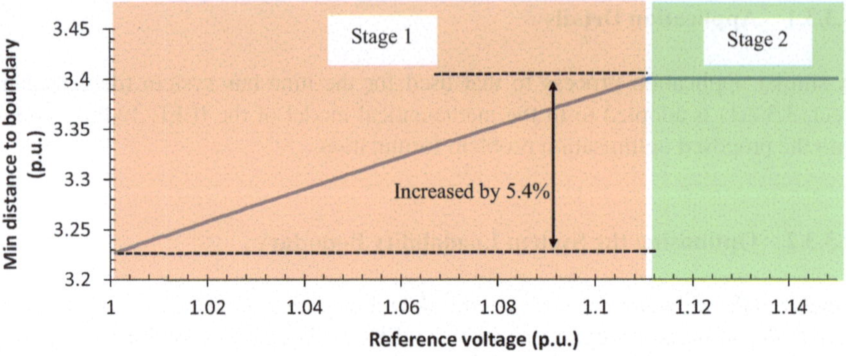

Fig. 3.14 System closest loadability point with the varying reference voltage, IEEE 39-bus system. Adapted from Chen et al. [10], released under a CC BY 4.0 license; https://doi.org/10.1016/j.ijepes.2021.107182

margin remains constant, as the busbar voltage limit is encountered. The reference voltage of 1.111 p.u. is the optimal value obtained using the continuation method. The optimal reference voltage calculated from the proposed optimisation-based framework is 1.110 p.u., which is within the acceptable relative error tolerance of 10^{-2}.

The results within this section have demonstrated not only the effectiveness of the proposed optimisation-based approach in improving the system's loadability, but also the applicability of the developed methodology in a power large system.

3.4 Summary

In this chapter, a new optimisation-based framework has been developed to maximise the closest maximum loadability point of a system, with the simultaneous consideration of SNB, LIB, and HB. The effectiveness of the proposed methodology has been demonstrated in various test networks, including a single-machine three-bus system, a multi-machine nine-bus system, and a large, IEEE 39-bus test system. It has been clearly shown that the proposed analysis framework can be used to assess and subsequently optimise system loadability. The framework is as accurate as the continuation method and has the ability to improve overall system loadability in the given parameter space.

The system loadability concept is renewed by considering SNB, LIB, and HB simultaneously. Based on this revised concept, five loadability scenarios have been discussed and are illustrated in three- and nine-bus systems. It is evident that when SNB, HB, and LIB are considered simultaneously, the system's loadability boundary consists of different bifurcations. This necessitates the development of a new optimisation problem formulation to analyse the updated system loadability. A comparative study of system loadability boundaries with different types of bifurcation shows that the boundary based on a single type of bifurcation is significantly different from a boundary consisting of multiple, simultaneous bifurcations. This highlights the importance of the simultaneous consideration of all relevant bifurcations when performing loadability assessment. An analysis that considers only one type of bifurcation might lead to an erroneous assessment of the system's loadability.

The proposed multi-stage framework consists of three optimisation loops (the inner, middle, and outer loops). The function of the inner loop is to obtain the system's maximum loadability point when considering simultaneous multiple bifurcations. The system loadability boundary is then produced by connecting the obtained system loadability points. The shortest loading margin, which is calculated from the middle loop, is then fed into the outer loop. Finally, the outer loop determines the optimal parameter to optimise the system's shortest loading margin in a parameter space and hence improve the system's overall loadability. The continuation method is adopted to verify the results from the new optimisation-based method. The studies performed in various test networks have clearly demonstrated the effectiveness of the proposed multi-stage optimisation-based analysis framework. It is clear that the developed multi-stage framework can perform system loadability assessments accurately. Furthermore, by applying the proposed optimisation method, the system's shortest loading margins can be improved by 9.9%, 8.4%, and 5.4% for three-bus, nine-bus, and IEEE 39-bus systems, respectively. Power system operators can employ the developed framework to accurately identify the system's loadability boundary with the simultaneous consideration of multiple bifurcations and to improve the system's overall loadability margin.

This chapter has demonstrated the necessity to consider multiple types of bifurcation simultaneously in system loadability assessment. The results of the analysis show that bifurcation-based theory is suitable for analysing the small-disturbance voltage stability of a system. However, applying bifurcation theory to investigate other types of stability remains unexplored. The next chapter uses a bifurcation-based approach to assess the system's small-disturbance stability.

References

1. Kabouris J, Vournas CD, Efstathiou S, Manos GA, Contaxis GC (2002) Voltage security considerations in an open power market. In: DRPT2000. International conference on electric utility deregulation and restructuring and power technologies. proceedings (Cat. No.00EX382), pp 278–283
2. Pasrija AK, Bahar LY (1986) Static bifurcations in electric power networks: loss of steady-state stability and voltage collapse. IEEE Trans Circuits Syst 33(10):981–991
3. Chiang HD, Wang CS, Flueck AJ (1997) Look-ahead voltage and load margin contingency selection functions for large-scale power systems. IEEE Trans Power Syst 12(1):173–180

4. Dobson I, Lu L (1992) Voltage collapse precipitated by the immediate change in stability when generator reactive power limits are encountered. IEEE Trans Circ Syst I Fundam Theory Appl 39(9):762–766
5. Van Cutsem T, Vournas CD (2007) Voltage stability of electric power system, 1st ed. Springer
6. Van Cutsem T, Kabouris J, Christoforidis G, Vournas CD (2005) Application of real-time voltage security assessment to the Hellenic interconnected system. IEE Proc: Generation, Transm Distrib 152(1):123–131
7. Cañizares CA, Hranilovic S (1994) Transcritical and Hopf Bifurcations in AC/DC Systems. In: Proceeding bulk power system voltage phenomena—III seminar on voltage stability, security and control, pp 105–114
8. Makarov YV, Dong ZY, Hill DJ (2002) A general method for small signal stability analysis [of power systems]. In: 20th international conference on power industry computer applications, pp 280–286
9. Hill DJ, Dong ZY (2000) Nonlinear computation and control for small disturbance stability. IEEE Power Eng Soc Transm Distrib Conf 2:836–841
10. Chen Y, Preece R, Barnes M (2021) Improving system loadability with the consideration of multiple bifurcations. Int J Electr Power Energy Syst 132:107182
11. Avalos RJ, Cañizares CA, Milano F, Conejo AJ (2009) Equivalency of continuation and optimization methods to determine saddle-node and limit-induced bifurcations in power systems. IEEE Trans Circuits Syst I Regul Pap 56(1):210–223
12. Karystianos ME, Maratos NG, Vournas CD (2007) Maximizing power-system loadability in the presence of multiple binding complementarity constraints. IEEE Trans Circuits Syst I Regul Pap 54(8):1775–1787
13. Guckenheimer J, Holmes P (1983) Nonlinear oscillations, dynamical systems, and bifurcations of vector fields, 1st edn. Springer, New York, NY
14. Jalali A, Aldeen M (2018) Efficient method to identify saddle-node and limit-induced bifurcation points of power system. IET Gener Transm Distrib 12(17):4034–4044
15. Venkatasubramanian V, Schättler H, Zaborszky J (1995) Dynamics of large constrained nonlinear systems-a taxonomy theory. Proc IEEE 83(11):1530–1561
16. Vournas CD, Karystianos M, Maratos NG (2000) Bifurcation points and loadability limits as solutions of constrained optimization problems. In: IEEE power engineering society transmission and distribution conference, pp 1883–1888
17. Zimmerman RD, Murillo-s CE (2020) Matpower user's manual 7.1
18. Dobson I, Lu L (1993) New methods for computing a closest saddle node bifurcation and worst case load power margin for voltage collapse. IEEE Trans Power Syst 8(3):905–913
19. Li P, Qi J, Wang J, Wei H, Bai X, Qiu F (2017) An SQP method combined with gradient sampling for small-signal stability constrained OPF. IEEE Trans Power Syst 32(3):2372–2381
20. Nocedal J, Wright S (2006) Numerical optimization, 2nd edn. Springer, New York
21. Schittkowski K (1986) NLPQL: a Fortran subroutine for solving constrained nonlinear programming problems. Artic Ann Oper Res 5(2):485–500
22. Lehmköster C (2002) Security constrained optimal power flow for an economical operation of FACTS-devices in liberalized energy markets. IEEE Trans Power Deliv **17**(2), 603–608
23. Fletcher R, Powell MJD (1963) A rapidly convergent descent method for minimization. Comput J 6(2):163–168
24. Goldfarb D (1970) A family of variable-metric methods derived by variational means. Math Comput 24(109):23–23
25. MATLAB – MathWorks—MATLAB & Simulink, The MathWorks, Inc, 2018. [Online]. Available: https://uk.mathworks.com/products/matlab.html. [Accessed: 10-Mar-2021]
26. Vlachogiannis JG, Lee KY (2006) A comparative study on particle swarm optimization for optimal steady-state performance of power systems. IEEE Trans. Power Syst 21(4):1718–1728
27. Sui X, Tang Y, He H, Wen J (2014) Energy-storage-based low-frequency oscillation damping control using particle swarm optimization and heuristic dynamic programming. IEEE Trans Power Syst 29(5):2539–2548

28. Canizares CA (1998) Calculating optimal system parameters to maximize the distance to saddle-node bifurcations. IEEE Trans Circ Syst I Fundam Theory Appl 45(3):225–237
29. Zhou Y, Ajjarapu V (2005) A fast algorithm for identification and tracing of voltage and oscillatory stability margin boundaries. Proc IEEE 93(5):934–946
30. Greene S, Dobson I, Alvarado FL (1997) Sensitivity of the loading margin to voltage collapse with respect to arbitrary parameters. IEEE Trans Power Syst 12(1):262–272
31. Dobson I, Lu L (1992) Computing an optimum direction in control space to avoid saddle node bifurcation and voltage collapse in electric power systems. IEEE Trans Automat Contr 37(10):1616–1620
32. (Prabha) Kundur P, Balu NJ, Lauby MG (1994) Power system stability and control, 1st ed. McGraw-Hill
33. Tan CW, Varghese M, Varaiya P, Wu FF (1995) Bifurcation, chaos, and voltage collapse in power systems. Proc IEEE 83(11):1484–1496

Chapter 4
Analysis of Feasibility Region Boundaries with Consideration of Multiple VSCs

It is anticipated that a large number of VSCs will be integrated into future power systems, which may potentially be detrimental to system stability. Alongside this large-scale adoption of VSCs, it is also predicted that multiple VSCs may be connected close together, or even that multiple VSCs will be integrated into the network at the same point. As the electrical distance between such VSCs will thus be small, and considering the fact that the control loops of various VSCs will most likely be designed to work at similar speeds, the potential exists for unwanted interactions among the control loops of different devices. Chapter 3 has demonstrated the ability of bifurcation theory to analyse the voltage stability of a system. The bifurcation theory-based approach will be adopted in this chapter to analyse the newly emergent VSC-related small-disturbance instabilities.

Many papers have performed research into grid-connected VSC-associated stability issues, using various analysis techniques, including modal analysis [1], impedance-based Nyquist criterion [2], and so on. A review of this work has been presented in Sect. 1.5.1. Although there exist numerous instability studies in systems with multiple VSCs connected, there is a need for a methodology that can determine the system feasibility region boundary in a multi-dimensional (i.e., N-dimensional) domain and for the completion of a more comprehensive study of the impact on a system feasibility region with N different system parameters varying simultaneously.

As identified in the research gap in Sect. 1.5.4, the sensitivity analysis of the system feasibility boundary performed in previous work has typically considered the variation in only one specific parameter at a time, with other parameter values fixed at nominal values. Any such sensitivity result is, therefore, considered local as it ignores the effects of the interactions between parameters, and higher-order cross-effects, along the system feasibility boundary. There is a need for a sensitivity

Y. Chen, *Stability Assessment of Power Systems with Multiple Voltage Source Converters*, Springer Theses, https://doi.org/10.1007/978-3-031-63095-8_4

analysis that provides global parameter sensitivity over a given parameter range while simultaneously considering the effects of interactions between parameters.

This chapter presents the second contribution of this thesis, a methodology to efficiently identify a system's multi-dimensional feasibility region boundary with respect to small-disturbance stability in cases where multiple parameters may vary simultaneously and multiple types of bifurcations are considered. The numerical method, i.e. quasi-Monte Carlo simulation of a wide range of system operating scenarios, is adopted within the developed methodology, with quasi-random sampling used to determine the system's initial conditions in an N-dimensional space. For each computational scenario, the continuation method is implemented to determine the system feasibility points. A data-driven model is then created based on the interpolation of results from the continuation method and used to efficiently identify the system feasibility region boundary in N-domains. The developed data-driven model can then be adopted in the further sensitivity analysis.

In addition to the developed methodology, this section also proposes a numerical analysis framework for sensitivity studies based on the data-driven model to analyse the system feasibility boundary using the Spearman correlation coefficient. This represents the third contribution of this thesis. By applying Spearman correlation analysis, this framework considers the interdependency among various parameters and sensitivity across the entire parameter range. Parameters that have significant impacts on the system feasibility region can then be identified by examining the correlation index values. Additionally, based on the assessment of the Spearman correlation coefficient, the proposed analysis framework can be used to identify interactions between the VSCs and the network.

A small test network consisting of two VSCs is initially used to verify the effectiveness of the developed methodology and perform sensitivity studies. An illustrative example on a IEEE 39-bus power system considering multiple VSCs is then included to further confirm the applicability of the developed methodology and analysis framework in larger power systems.

4.1 Background

This section introduces the concepts and analysis techniques used in this chapter, including the feasibility region, quasi-random sampling, interpolation technique, and correlation coefficients.

4.1.1 The Feasibility Region

This section introduces the feasibility region concept. Using the definition given in this section, a methodology will be developed in Sect. 4.2 to determine the feasibility region of systems. Consider a system equilibrium point calculated using (2.65). By varying the system bifurcation parameter μ, the system dynamics can be altered, as the system equilibrium point moves from one location to another. Based on the work in [4], if the topological structure of the system remains unchanged under the influence of a small disturbance, the system can be considered to be structurally stable. The typal region initially proposed in [4] is defined as the system parameter space in which the system is structurally stable. Theoretically, the typal region offers a comprehensive understanding of critical dynamic changes in the system, as it incorporates both local and global bifurcations. However, these global bifurcations are dense in an open set of parameter spaces, making it extremely difficult to analyse system stability in large systems. Consequently, [5] proposed the use of feasibility regions developed from typal regions. The feasibility region is a union of certain typal regions where the system operation point is locally stable under quasi-static parameter variations. The boundary of the feasibility region is thus a composite of a set of closures of three types of codimension-one bifurcation surfaces (i.e. SNB, HB, and SIB). The concept of the feasibility region developed in [5, 6] is adopted in this thesis. The mathematical expression for the feasibility boundary that surrounds the feasibility region is shown in (4.1), and this is valid for the codimension-one bifurcation surfaces defined in [5, 6]. The term codimension-one bifurcation surface itself indicates that the surface roughly consists of $(N - 1)$ dimensions when an N-dimensional parameter space is considered.

$$\Omega = \{\mathrm{SNB} \cup \mathrm{HB} \cup \mathrm{SIB}\} \tag{4.1}$$

It has been stated in Chap. 2 that this thesis is not interested in SIB, as there is a lack of real power system collapse events caused by SIB. Hence, the corresponding mathematical expression for the feasibility boundary discussed in this chapter is shown in (4.2).

$$\Omega = \{\mathrm{SNB} \cup \mathrm{HB}\} \tag{4.2}$$

4.1.2 *Quasi-Random Sampling*

The methodology developed in this chapter to determine the multiple-dimensional system feasibility region boundary is based on quasi-Monte Carlo simulations. Specifically, this methodology identifies the system feasibility region boundary by numerically computing a large number of operating points located on the feasibility hypersurface. For such a quasi-Monte Carlo simulation, one of the key steps is to initialise the system in N-dimensional parameter space. As N increases, it is computationally exhaustive to manually set up all possible initial system conditions. Consequently, quasi-random sampling is adopted. Within the proposed methodology, the quasi-random sampling technique is used to determine the vectors of the change of bifurcation parameters, which, if followed, will generate a series of operating points on the system feasibility boundary. The details of the application of quasi-random sampling are included in Sect. 4.2.2. Compared with random numbers, quasi-random sets can be used to fill the search space more uniformly [7]. Quasi-random sequences are thus often applied in the quasi-Monte Carlo method to improve convergence properties when a multi-dimensional system is used [7]. The proposed methodology adopts quasi-random sampling based on Halton sequences. Halton sequences are low-discrepancy sequences and are represented as a natural generalisation of Van der Corput sequences to higher dimensions [7, 8]. Halton sequences can be generated by (4.3)–(4.5).

Consider $n \in N_0 \rightarrow [0,\ 1)$ has a unique digital expansion in base b, as presented in (4.3) [9].

$$n = \sum_{i=0}^{m} a_i b^{i-1} \tag{4.3}$$

In (4.3), a_i references the non-negative integers less than b, while b is the base value (also known as the prime), which is an integer value greater than 1; $m = \log_b n$.

The radical inverse function $\varphi_b(n)$ can be defined as in (4.4).

$$\varphi_b(n) = \sum_{i=0}^{m} \frac{a_i}{b^{i+1}} \tag{4.4}$$

On the basis of the Van der Corput sequence, (4.5) thus provides the definition of the Halton sequence X_n in $[0,\ 1)$.

$$X_n = (\varphi_{b1}(n), \varphi_{b2}(n) \ldots \varphi_{bn}(n)) \tag{4.5}$$

In (4.5), $b_1, b_2 \ldots b_n$ are the integer numbers greater than 1 and are known as pair-wise co-primes.

The Halton sequences can be viewed as an approach to distributing points over the space in a highly uniform manner by using different prime bases in each dimension. However, due to the way in which quasi-random sequences are generated, a problem arises from the correlation between the radical inverse functions for different dimensions in Halton sequences. This undesirable correlation becomes significant in systems with higher dimensions. To improve the uniformity of the Halton sequences, part of the quasi-random points skip and leap over, or scramble values in the sequence to break the correlation among dimensions [10].

Figure 4.1 shows the distribution of sampling points in a two-dimensional space using (a) quasi-random variables generated by the Halton sequence and (b) random variables. Each diagram contains 100 sampling points. Within Fig. 4.1b, two oval regions that have the same area are surrounded by the dashed and solid curves respectively. In an ideal case in which sampling points are evenly distributed over the search space, each area should contain the same number of sampling points. However, Fig. 4.1b shows that there are few points in the area surrounded by the dashed curve

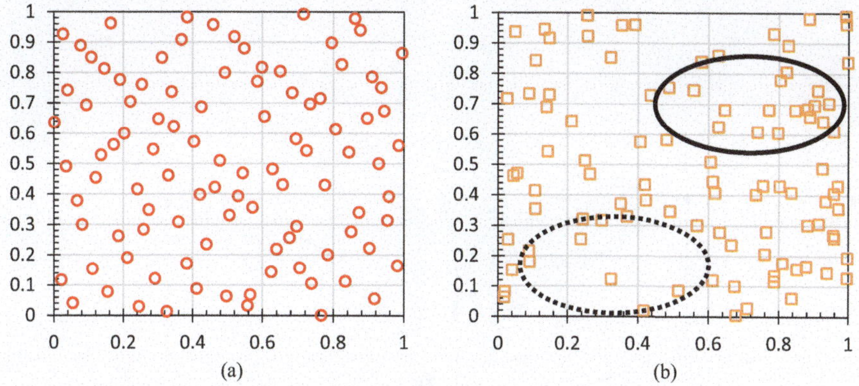

Fig. 4.1 Comparison of **a** Halton sequence sampling and **b** random sampling

and more points squeezed in the area surrounded by the solid curve. Compared with random sampling, the sampling points generated by the Halton sequence can fill the search space more evenly, as is shown in Fig. 4.1a. Based on this, the quasi-random sampling based on Halton sequences has been employed in the developed methodology to thoroughly compute the system feasibility region boundary.

Within the work described in this chapter, a quasi-random sampling technique based on the Halton sequence was applied using the *haltonset* function in MATLAB® 2019.

4.1.3 The Interpolation Technique

The methodology developed in this chapter involves computing a large number of operating points on the feasibility region boundary. Therefore, the stopping criterion will terminate the simulation when a sufficient amount of data has been obtained to represent the system feasibility boundary (for use in further studies). In general, a rule of thumb within the proposed approach is to terminate the simulation when the estimated feasibility region boundary formed by the obtained data can produce suitably accurate results as those computed from the actual feasibility hypersurface. Interpolation is a technique used to predict a new data point from a set of known data points. Hence, the interpolation technique is employed in the proposed methodology to form a data-driven model that represents the feasibility region based on the currently obtained data. The stopping criterion based on the interpolation technique is discussed in Sect. 4.2.4. Apart from the application in the stopping criterion, the data-driven model will be used as an efficient tool to compute critical points on the feasibility region boundaries in Sects. 4.3–4.4, where it is used for in-depth sensitivity analysis.

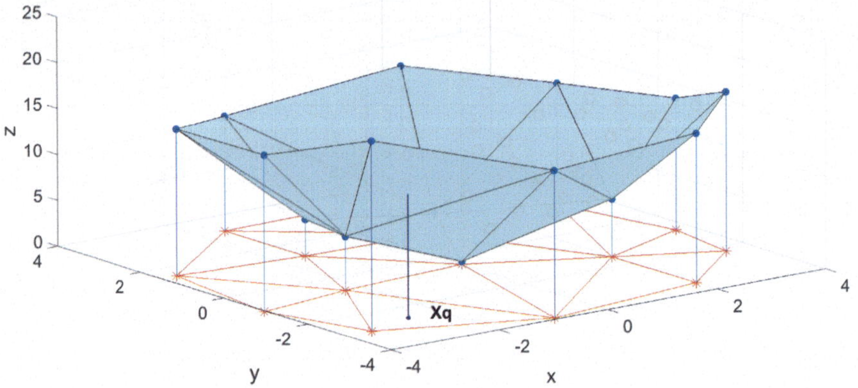

Fig. 4.2 Example of the interpolation of a paraboloid

Figure 4.2 represents an example of interpolation. The round blue points in Fig. 4.2 represent a data set on the surface of a paraboloid in the *x*-*y*-*z* axes, and the red star points are the projections of the round points on the *x*-*y* plane. A Delaunay triangulation is completed by connecting the adjacent three star points in the *x*-*y* plane [11]. The vertices of the triangulated points can then be lifted by the magnitude of their value on the *z*-axis into a dimension orthogonal to the *x*-*y* plane to form an interpolated paraboloid surface on the *x*-*y*-*z* axes. The interpolated surface is coloured light blue in Fig. 4.2. Finally, the value of the interpolation is evaluated at the query point X_q, based on the interpolated paraboloid surface. This step involves traversing the triangulation data structure to identify the triangle that encloses the query point. Once this triangle is determined, a linear algorithm is adopted in this chapter to calculate the value at the given query point [12]. Specifically, this algorithm generates linear polynomials using three vertices to fit the triangle formed by those three points.

Interpolation is adopted in this chapter using the *griddatan* function in MATLAB® 2019. The *griddatan* function can be used for N-dimensional scatter-data interpolation. It will first estimate the hypersurface formed by the current data. Then it will interpolate the surface at given query points to return the estimated values at those query points.

4.1.4 The Correlation Coefficient

The proposed analysis framework adopts a statistical approach, using correlation techniques to analyse the obtained data which represents the system feasibility boundary. The sensitivity analysis of the multiple-dimensional feasibility region of the system based on the correlation coefficient will be discussed in Sects. 4.3–4.4.

Interdependency indicates the extent to which variables are related. As the system feasibility region boundary is formed by the test parameters, those having a high dependency are considered as key parameters. Numerous correlation techniques exist which provide quantitative assessments of the interdependency between variables. Based on the type of given data, correlation analysis tools can be classified into two types: parametric and nonparametric. This section introduces two typical correlation techniques: the Pearson correlation (parametric) [13] and the Spearman correlation (nonparametric) [14].

4.1.4.1 The Pearson Correlation Coefficient

The Pearson correlation is a parametric correlation coefficient that provides a quantitative measure of the dependency between variables. Assuming there are two variables, X and Y, the definition of the Pearson correlation index is as shown in (4.6).

$$\rho_{X,Y} = \frac{\mathrm{cov}(X, Y)}{\sigma_X \sigma_Y} \tag{4.6}$$

In (4.6), $\mathrm{cov}(X, Y)$ is the covariance of the variables, while σ_X and σ_Y represent the standard deviations of the corresponding variables.

4.1.4.2 The Spearman Correlation Coefficient

The Pearson correlation is a measure of the linear relationships among random variables, whereas the Spearman correlation, as a measure of nonparametric correlation, assesses the rank dependency between variables, regardless of whether their correlations are linear or nonlinear [14]. As discussed in Chap. 2, a power system is a highly nonlinear system. Therefore, the Spearman correlation best fits the scope of this section. In general, the Spearman correlation coefficient is the Pearson correlation coefficient of the rank variables. For two sets of data, X and Y, , the Spearman correlation coefficient is calculated as in (4.7).

$$r_s = \rho_{R(X),R(Y)} = \frac{\mathrm{cov}(R(X), R(Y))}{\sigma_{R(X)} \sigma_{R(Y)}} \tag{4.7}$$

In (4.7), $R(X)$, $R(Y)$ is the rank of corresponding variables X and Y. The rank of the variable is the order number of the variable when all variable components in a set are sorted in value-increasing order.

The Spearman correlation index varies between $-\ 1$ and $+\ 1$. A value of $+\ 1$ means that X and Y are monotonic with a positive linear rank correlation. A value of $-\ 1$ also represents a monotonic relationship, but with a negative rank correlation. A value of zero means there is no rank correlation.

4.1.4.3 The Partial Correlation Coefficient

Within the multi-dimensional sensitivity study performed in this work, the standard Spearman correlation coefficient will tend to under-represent the relationships between parameters, as it will be diluted by variations caused by other parameter changes. Therefore, the partial Spearman correlation coefficient is adopted in this thesis, where the correlation index between a selected pair of bifurcation parameters is calculated while controlling for variations in the remaining considered bifurcation parameters [15], [16]. Considering three different variables, X, Y, and Z, the partial correlation coefficient between X and Y when the effects of Z have been cancelled is given by (4.8).

$$\rho_{X,Y.Z} = \rho\left(X - \hat{X}, Y - \hat{Y}\right) \tag{4.8}$$

In (4.8), $\hat{X}$ and $\hat{Y}$ are the predictions of the linear model, which represent the correlations of X and Y with respect to the other parameter, Z.

4.2 Identification of System Feasibility Region Boundary

This section introduces the methodology developed to efficiently identify a system N-dimensional feasibility region based on the continuation method. Unlike previous work, in which the system stability boundary has been directly represented by sets of mathematical equations, this section adopts a numerical method to identify the points on the feasibility region boundary of the system in N-domains. In general, the developed methodology involves quasi-Monte Carlo simulations of a large number of system scenarios, in which the initial conditions of the system are generated based on a quasi-random sampling technique (i.e. the Halton sequence). Following the creation of each initial condition, a system stability point that is on the feasibility boundary can be obtained using the continuation method. The interpolation technique is then adopted to develop a data-driven model which approximates the real system feasibility hypersurface based on the currently obtained system stability points. The system will continuously identify system stability points until the stopping criterion is met. The stopping criterion will terminate the simulation when the new result generated using the continuation method no longer adds new information to the interpolated feasibility region boundary.

4.2.1 *The Continuation Method*

The continuation method developed in Chap. 2 will be employed in this chapter to trace the system equilibrium trajectory and determine the critical points, i.e., various bifurcation points. It should be noted that this chapter represents the second contribution: a methodology to identify system multi-dimensional feasibility region boundary, which fills a research gap identified in Sect. 1.5.4. This is different from Chap. 3, in which most effort is invested into developing a new optimisation problem formulation that could consider multiple types of bifurcations simultaneously. Additionally, the continuation method provides the most detailed information about system stability along the equilibrium trajectory. Consequently, the continuation method is used in this chapter to compute the system's critical points. The application of other time-efficient techniques, for example, the optimisation-based approach, will not affect the results obtained. However, those techniques can be used to reduce simulation time, if desired.

4.2.2 The Direction of Change of the Bifurcation Parameter

The concept of the direction of increase was briefly introduced in Chap. 2. However, most of the effort so far has been invested into its application within two dimensions where the system's active and reactive power loading are considered as bifurcation parameters. This section further extends this concept to N dimensions, considering situations where any system parameters can be selected as bifurcation parameters.

When there is only one bifurcation parameter, the change direction of this bifurcation parameter must be within a one-dimensional space, i.e. either increasing or decreasing. However, when multiple bifurcation parameters are considered, the direction of change of the bifurcation parameter group becomes a multi-dimensional vector, which is more complex. Consider the case shown in Fig. 4.3: two bifurcation parameters, X and Y, change simultaneously at their own rates, represented by the two vectors k_1 and k_2, respectively. Defining a standardised bifurcation parameter, λ, the changing rate of each bifurcation parameter X and Y is the derivation of each bifurcation parameter with respect to λ, i.e. $k_1 = \frac{dX}{d\lambda}$, and $k_2 = \frac{dY}{d\lambda}$. The superposition of these two vectors will then form a new vector, $K = k_1 + k_2$, which is the direction of change of the bifurcation parameter group overall. Following this direction of change, K, a critical point is reached by tracing the system equilibrium curve with the continuation method. Critical points are obtained for each established direction of change of the bifurcation parameter group. A system feasibility boundary is created when these critical points are connected, represented by the grey line in Fig. 4.3, and this surrounds the feasibility region. A similar theory can be applied to cases with higher dimensions where N > 2. The two-dimensional case can be generalised to an N-dimensional case where the direction of increase is an N-dimensional vector.

As N increases, it becomes an extremely computationally intensive process to determine every possible direction of change within the proposed approach. The proposed methodology selects a direction of change based on quasi-random sampling. Consider the case shown in Fig. 4.3, where $N = 2$. The possible directions of change of the bifurcation parameter group are represented by the red unit

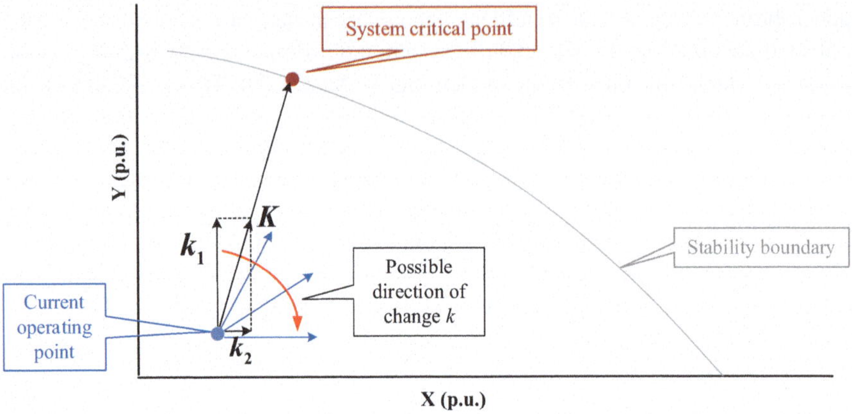

Fig. 4.3 Example of the direction of change in a bifurcation parameter group. © 2023 IEEE, reprinted, with permission, from Chen et al. [3]

quarter-circle in the diagram. A vector pointing from the current operating point to a point on that quarter-circle represents a certain direction of change of the bifurcation parameter group. The Halton sequence discussed in Sect. 4.1.2 is adopted to generate a series of quasi-random sampling points that are uniformly distributed over the quarter-circle [9]. This subsequently establishes a series of directions of change of the bifurcation parameter group, which will be evaluated to establish the feasibility region boundary and, therefore, identify the feasibility region.

4.2.3 Interpolated Feasibility Boundary

The interpolation technique introduced in Sect. 4.1.3 will be used to approximate the system's stability hypersurface formed by the previously obtained, quasi-randomly sampled stability points. Combined with the results from the continuation method, the interpolated feasibility hypersurface generated is incorporated into the stopping criterion to terminate the simulation when sufficient quasi-random samples have been collected.

4.2.4 The Stopping Criterion

With a quasi-random sampling approach, a stopping criterion is needed to terminate the numerical simulation process at an appropriate time. The stopping criterion ends the simulation if the average relative difference between the outcomes from the data-driven prediction model (from the interpolated feasibility boundary) and actual results (from the continuation method) with respect to the actual results is smaller than a previously set threshold (ε). This process is represented in mathematical form in (4.9).

$$\frac{1}{n}\sum_{i}^{n}\left|\frac{f_{\text{actual}}(K_i)-f_{\text{interpolate}}(K_i)}{f_{\text{actual}}(K_i)}\right|<\varepsilon$$

$$\text{over the last } n \text{ points} \tag{4.9}$$

In (4.9), K is the direction of the increase of bifurcation parameters; $f_{\text{interpolate}}(K_i)$ is the result using the interpolation method introduced in the previous section; and $f_{\text{actual}}(K_i)$ is the result from the actual model, which is obtained by running the continuation method.

Based on (4.9), the stopping criterion will stop the simulation when any additionally obtained points are close to or on the predicted linear stability hypersurface formed by the previous stability points. That is to say, the simulation will be terminated when additional stability points no longer add new information to the system stability hypersurface. The system feasibility region boundary is obtained based on the continuation method until the stopping criterion is met. After that, analysis is applied based on the results from the interpolated hypersurface generated by the *griddatan* function in MATLAB®.

The required number of sampled stability points using the continuation method with respect to different combinations of n and ε are presented in Table 4.1. This comparative study is completed in a seven-dimensional search space, which is formed by parameters from one VSC and the network. The table indicates that a higher value of n and smaller ε will offer higher accuracy. However, this requires more directions of change of the bifurcation parameter group and more system stability points to be obtained from the continuation method. Besides, a smaller relative error threshold results in a smaller step-size setting in the continuation method, which ultimately leads to an increased computational burden. Considering simulation time and the relative error threshold setting of the continuation method in Chap. 2, $n = 5$, $\varepsilon = 0.01$ is used in this section to provide the illustrative results.

Table 4.1 Number of simulations required for different sets of parameters (n, ε) in the stopping criterion. © 2023 IEEE, reprinted, with permission, from Chen et al. [3]

Number of points used in the stopping rule (n)	Number of simulations to achieve the desired accuracy		
	$\varepsilon = 0.1$	$\varepsilon = 0.01$	$\varepsilon = 0.001$
2	101	162	2934
3	101	161	6863
4	107	1198	10,032
5	107	2019	56,732

A secondary stopping criterion enforced is the maximum number of simulations considered. To limit the overall simulation time to within 24 h for this study (using the computational resource available for this research), the maximum number of simulations is set to 10^7.

4.2.5 *Detailed Steps*

The detailed steps for applying the proposed methodology are as follows:

Step 1: Run the first 100 directions of change generated via the Halton sequence. For each direction of change, one system critical point (representing a stability limit) will be obtained using the continuation method. Thus, 100 system critical points will be obtained. These 100 points act as the minimum number of points needed before attempting to stop the simulation using the stopping criteria.

Step 2: Use the interpolation technique to approximate the stability hypersurface formed by all currently obtained critical points.

Step 3: Run an additional n directions of changes of the bifurcation parameter group (in this work, $n = 5$, as is discussed in the previous section) using the continuation method.

Step 4: Use the interpolated stability hypersurface to estimate the system critical point values for these n points, and determine whether the stopping criterion given in (4.9) is met. If yes, stop the simulation. If not, add these n points to the dataset of system critical points and repeat Steps 2 and 3 until one of the stopping criteria is met.

These steps will produce the system feasibility region when the simulation process ends.

4.3 Feasibility Analysis of a Small Test Network

The previous sections developed a methodology to efficiently identify the system's N-dimensional feasibility region. This section applies the developed approach to a small test network with multiple VSCs connected. The feasibility region formed when considering the parameters of one VSC, and the parameters of two VSCs, will be investigated as two separate case studies. The validity of the results obtained from the proposed methodology will be demonstrated using time-domain simulations of the model developed within MATLAB® Simulink® 2020.

4.3.1 Test Network Details

Figure 4.4 shows a typical small power system model for analysing interactions among VSCs where two VSCs are mutually coupled by means of a small impedance [17, 18]. To take small-disturbance stability (specifically the converter-driven stability and the resonance stability in this chapter) into account, the dynamic network model is used with the consideration of network LC dynamics, which may potentially interact with the VSCs. That is to say, the dynamic network model developed in Sect. 2.1.7 is used in this chapter. The full system parameters and VSC control settings are given in Appendix A.4. The twelve different system parameters that form the considered feasibility region and their corresponding search ranges are shown in Table 4.2.

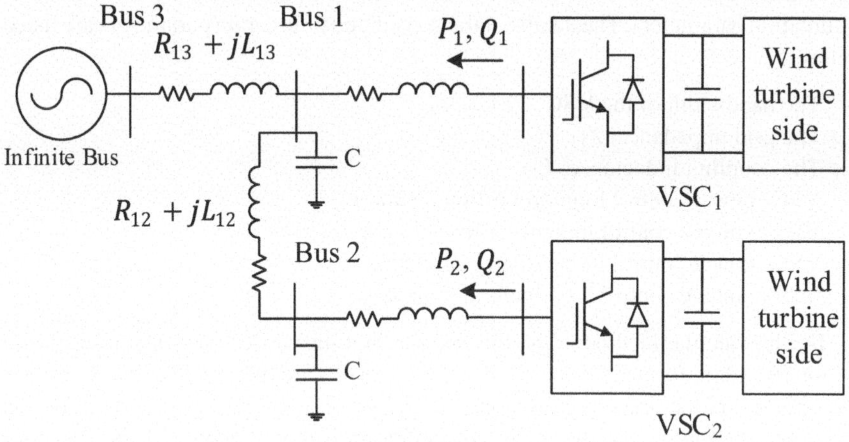

Fig. 4.4 Small-sized test system with multiple VSCs connected. © 2023 IEEE, reprinted, with permission, from Chen et al. [3]

Table 4.2 Key system parameter search ranges for a small test network. © 2023 IEEE, reprinted, with permission, from Chen et al. [3]

Key system parameter	System parameter search range
$\omega_{pll(1,2)}$ (rad/s)	[100, 1000]
L_{12} (p.u.)	[0.02, 0.2]
Z_{13} (p.u.)	[0.153, 1.53]
$k_{pP(1,2)}$ (p.u.)	[1, 10]
$k_{pV_{ac}(1,2)}$ (p.u.)	[0.2, 2]
$k_{iP(1,2)}$ (p.u.)	[120, 1200]
$k_{iV_{ac}(1,2)}$ (p.u.)	[215, 2150]

The initial values shown in Table 4.2 are tuned based on Skogestad internal model control (SIMC) [19], while the range of PI controller parameters in Table 4.2 offers an illustrative selection of possible value ranges. In this chapter, the range selected begins at the initial value and finishes at ten times the initial value. However, the methodology could theoretically be used for analysing any given range of PI controller parameters, if desired.

4.3.2 Parameters in One VSC

The methodology can be initially demonstrated by establishing the feasibility region when considering variations in the parameters of just one VSC. Therefore, the system parameters within one VSC and the overall network parameters are considered as bifurcation parameters. The details of the considered bifurcation parameters are listed below:

1. The bandwidth of the VSC_2 PLL: ω_{pll2}
2. The grid impedance: Z_{13}
3. The coupling inductance: L_{12}
4. VSC_2 power control loop proportional gain: k_{pP2}
5. VSC_2 voltage control loop proportional gain: $k_{pV_{ac}2}$
6. VSC_2 power control loop integral gain: k_{iP2}
7. VSC_2 voltage control loop integral gain: $k_{iV_{ac}2}$.

To illustrate the methodology, this section first introduces the system feasibility region boundary formed by $\omega_{pll2} - Z_{13}$. These results are then verified using another analysis technique, the eigenvalue trajectory. The feasibility region formed by the seven bifurcation parameters under consideration is then illustrated using a scatter plot. The partial Spearman correlation coefficient will be calculated to show the interdependency between bifurcation parameter pairs. Based on the calculated index, the interactions between control loops and influential parameters affecting the system feasibility region boundaries can be identified. Finally, in a similar manner to the examination of the boundaries formed by $\omega_{pll2} - Z_{13}$, the analysis procedure is applied to the parameter pair k_{pP2} and k_{pVac2}, which are positively correlated.

4.3.2.1 Boundaries in the $\omega_{pll2} - Z_{13}$ Plane

Firstly, the system feasibility boundaries (formed by $\omega_{pll2} - Z_{13}$) obtained from the interpolated hypersurface and from the full continuation method are shown in Fig. 4.5a. The results from these two methods are visually similar. This is guaranteed by the stopping criterion where the final five points generated from the hypersurface have an average relative error below 10^{-2} compared to the full continuation method. Taking the results obtained from the continuation method as the standard result, the relative root-mean-square deviation of the curve derived from the interpolated hypersurface is 5.56×10^{-3}. This is smaller than the relative error tolerance (10^{-2}) specified in the stopping rule, which illustrates the effectiveness of the proposed methodology with respect to producing feasibility boundaries accurately. All system feasibility boundaries in the following sections will be extracted from the interpolated hypersurface, with no further simulations carried out. Further analysis which, for brevity, is not shown has been performed to ensure that all relative errors between the interpolated hypersurface and the continuation method are below the threshold of 10^{-2}.

It should be noted that the feasibility boundary examined within this section is concerned with small-disturbance stability only, and other types of stability, such as frequency stability and transient stability, are neglected. The feasibility boundary derived in this section thus represents purely the small-disturbance stability boundary. If other types of stabilities are considered, this may result in other stability limits being met sooner, making this boundary potentially optimistic.

The remaining plots (b), (c), and (d) in Fig. 4.5 are produced by the interpolated hypersurface. With all other parameters set at the lower bound of the parameter search range, the feasibility region in each diagram is represented by the area below the corresponding curve. All curves in these charts reveal that the bandwidth of the PLL installed at VSC_2 has a negative relationship with the grid impedance in terms of the system feasibility region. Specifically, if the grid impedance increases, the PLL bandwidth must decrease to guarantee that the system can operate within a stable

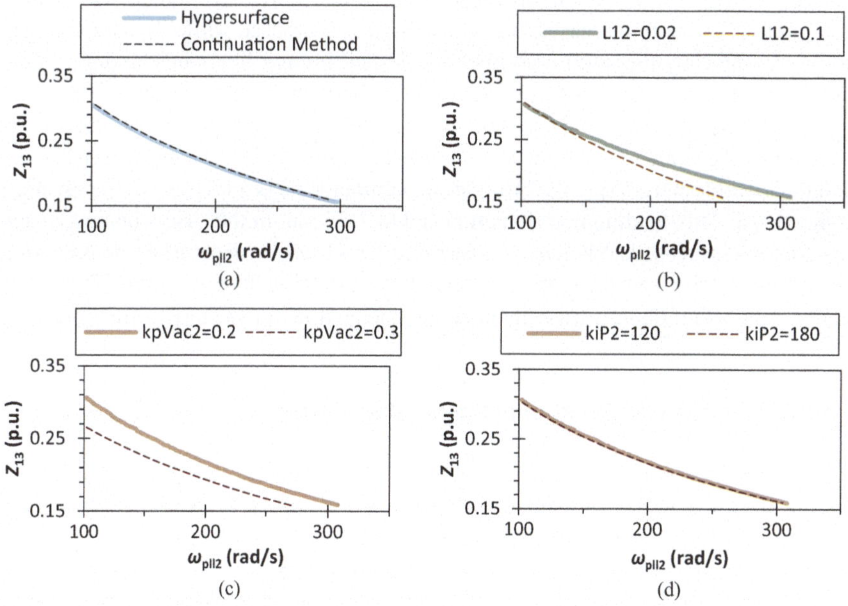

Fig. 4.5 Feasibility boundaries on the 2D plane of $\omega_{pll2} - Z_{13}$ showing **a** comparison of results from different approaches, **b** variation in L_{12}, **c** variation in $k_{pV_{ac}2}$, and **d** variation in k_{iP2} © 2023 IEEE, reprinted, with permission, from Chen et al. [3]

region. Moreover, the boundaries are also seen to be reduced if L_{12}, $k_{pV_{ac}2}$, and k_{iP2} are increased.

Figure 4.6a shows the key eigenvalue trajectory with respect to different parameter settings. As the feasibility region boundary in $\omega_{pll2} - Z_{13}$ is mainly formed by the HB type, the key eigenvalue is a pair of complex conjugated eigenvalues, and the trajectory of the eigenvalue is illustrated in Fig. 4.6a. The initial position of the eigenvalue in the complex plane is represented as a black point in Fig. 4.6a. If only ω_{pll2} is increased, the eigenvalue will move to the right-hand side of the complex plane through point A, following the solid blue trajectory. However, if the grid impedance also starts to increase, starting at the blue point, the key eigenvalue changes trajectory to follow the dashed orange line, and then passes through the imaginary axis at point B. Conversely, if the grid impedance Z_{13} is reduced from the blue point, the trajectory of the key eigenvalue will shift to the grey dotted line, and the eigenvalue will not cross the imaginary axis. The eigenvalue is more prone to instability (i.e., it has a more direct trajectory to the right-half plane) when both ω_{pll2} and Z_{13} are increased as compared to when only ω_{pll2} is increased. When ω_{pll2} is increased and Z_{13} is reduced, the critical eigenvalue remains on a stable trajectory. This confirms the negative correlation between ω_{pll2} and Z_{13} at the system feasibility boundary.

This relationship is further confirmed by considering the movement of the system operating point, as shown in Fig. 4.6b. The black points are critical points sampled using the proposed methodology, which together form the feasibility region boundary in $\omega_{pll2} - Z_{13}$, and the feasibility region is represented by the area below this boundary. The movements of the system operating point where only ω_{pll2} is increased, where ω_{pll2} and Z_{13} are both increased, and where ω_{pll2} is increased while Z_{13} is decreased, are represented by the solid blue line, the dashed orange line, and the dotted grey line, respectively (as previously). Figure 4.6b shows how these different operating condition trajectories drive the system closer to the feasibility region boundary or, in the case of the dotted grey trajectory, maintain a large stability margin. It is shown that the loading margin (i.e. the bifurcation parameter μ) is 3.08 p.u. when only ω_{pll2} is increased. This loading margin is reduced to 2.27 p.u. in the case where ω_{pll2} and Z_{13} are both increased. When ω_{pll2} is increased while Z_{13} is decreased, the operation point moves away from the feasibility boundary, following the grey curve. The results in Fig. 4.6 further demonstrate the negative correlation between ω_{pll2} and Z_{13}.

4.3.2.2 Considering Multi-dimensional Boundaries

The system feasibility boundaries in the $\omega_{pll2} - Z_{13}$ plane for two different values of L_{12} are shown in Fig. 4.5b, with the other parameters fixed. With further changes in L_{12}, different system feasibility region boundaries could be identified.

However, as previously discussed, exhaustively tracing all boundaries is not possible in multiple dimensions. Instead, the proposed methodology collects a data set of critical points (i.e. points along the feasibility region boundaries), which can be used to form the feasibility region boundary hypersurface. This can be demonstrated

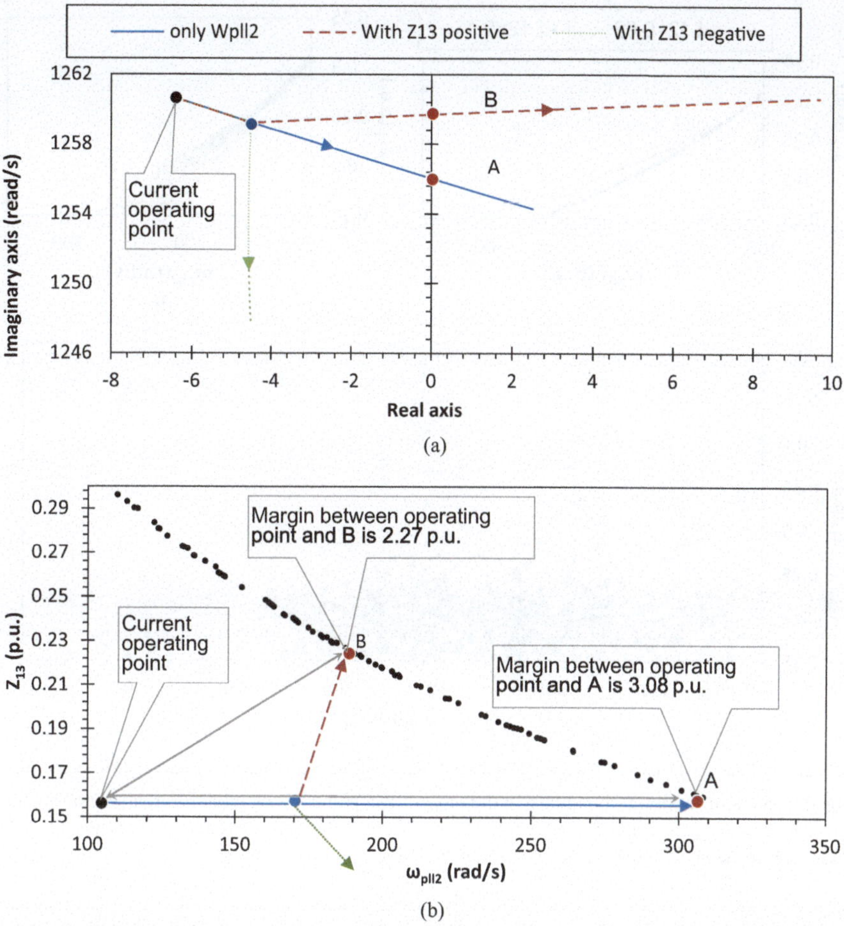

Fig. 4.6 Verification from eigenvalues with **a** the trajectory of key eigenvalue movement with respect to different parameter settings and **b** movement of the system's operating points from the current operating point towards the system feasibility region boundary on the $\omega_{pll2} - Z_{13}$ plane. © 2023 IEEE, reprinted, with permission, from Chen et al. [3]

by the system critical points in Fig. 4.7, which are plotted in the $\omega_{pll2} - Z_{13}$ coordinate space, as L_{12} is also varied. Each point in the scatter plot in Fig. 4.7a is produced using the specific values of L_{12} previously shown in Fig. 4.5b. These points produce the same feasibility boundaries as previously seen. For the scatter plot in Fig. 4.7b, L_{12} is varied further, across the full range of 0.02–0.2 p.u. Each point in Fig. 4.7b represents a point on the system's feasibility boundary hypersurface. Figure 4.7b can be viewed as a dimensionality reduction of the system's feasibility boundary in the $\omega_{pll2} - Z_{13} - L_{12}$ plane to the $\omega_{pll2} - Z_{13}$ plane. Figure 4.7c shows the system feasibility boundary in the $\omega_{pll2} - Z_{13}$ plane with the consideration of variations in

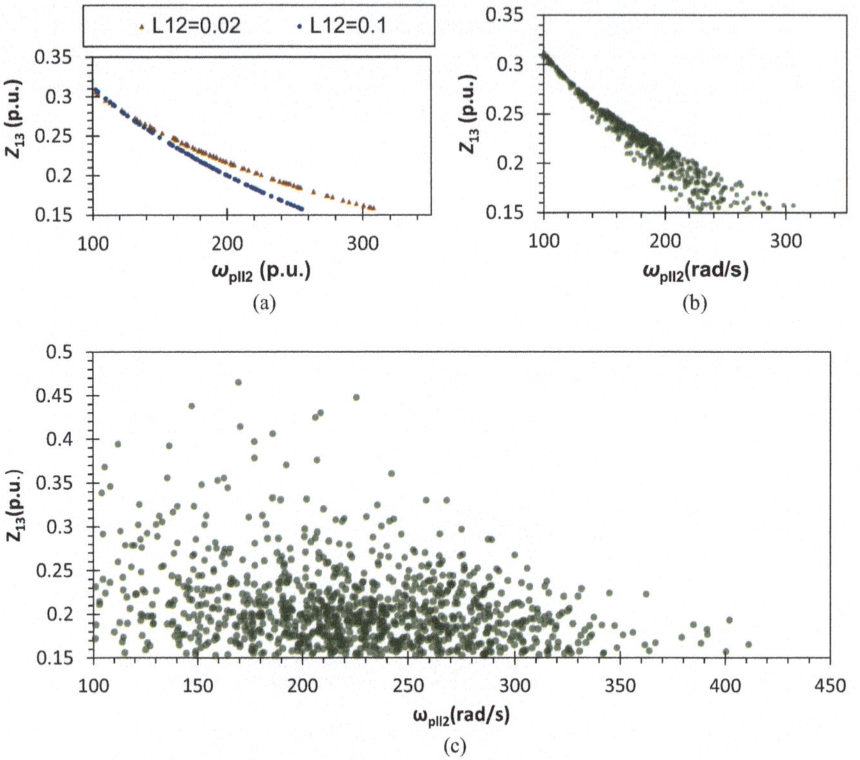

Fig. 4.7 System feasibility boundary in the $\omega_{pll2} - Z_{13}$ plane **a** with other bifurcation parameters fixed, **b** with ω_{pll2}, Z_{13} and L_{12} varying, and **c** with all bifurcation parameters varying. © 2023 IEEE, reprinted, with permission, from Chen et al. [3]

all seven parameters simultaneously. Hence, Fig. 4.7c is a dimensionality reduction of a seven-dimensional system feasibility boundary into the $\omega_{pll2} - Z_{13}$ plane.

4.3.2.3 Identifying Strong Parameter Relationships

It is hard to visualise and analyse a seven-dimensional system feasibility region directly, as the system stability boundary in this chapter consists of different types of bifurcation and cannot be represented by a standard mathematical model. Consequently, this chapter adopts a statistical method to identify the system stability limit points, which form the system feasibility region boundary, based on the continuation method and then analyse the obtained statistical data (i.e. Fig. 4.7c) using the partial Spearman correlation. The partial Spearman correlation coefficient of each pair of parameters is provided in Table 4.3. Each value represents the correlation index between parameters on the row and column, calculated using (4.8) in Sect. 4.1.4.3.

Table 4.3 indicates that the $\omega_{pll2} - Z_{13}$ partial correlation coefficient is – 0.71, a high absolute value. Such a large absolute correlation coefficient indicates that at the system feasibility region boundary, when ω_{pll2} is varied, Z_{13} is likely to vary significantly. Considering this from another perspective, both ω_{pll2} and Z_{13} are considered critical parameters, as they have a significant impact on the system feasibility region boundary formed by $\omega_{pll2} - Z_{13}$ when varied. Additionally, the negative sign indicates that an increase in PLL bandwidth ω_{pll2} will likely be accompanied by a reduction in the grid impedance Z_{13} at the feasibility region boundary. Viewing this from a system boundary perspective, if ω_{pll2} increases, then a lower value of Z_{13} is required to maintain the same margin to the system boundary. This aligns with previous discussions about the negative correlation between these parameters shown in Fig. 4.5.

Apart from the negative/positive correlation among parameters being identified by the sign of the partial Spearman correlation index, the value of the correlation index indicates the strength of the correlation between parameters. For example, as shown in Table 4.3, the absolute values of the partial Spearman correlation index between $k_{pV_{ac}2} - \omega_{pll2}$ (with a value of – 0.80) and $k_{pV_{ac}2} - Z_{13}$ (−0.68) are much larger than the absolute values for $L_{12} - k_{iP2}$ (−0.05) and $L_{12} - k_{iP2}$ (−0.01). This indicates that the relationship between parameters $k_{pV_{ac}2} - \omega_{pll2} - Z_{13}$ is stronger than that between $k_{iP2} - \omega_{pll2} - Z_{13}$. This can be confirmed by considering the system boundary curves previously presented in Fig. 4.5c and d, where $k_{pV_{ac}2}$ influences the system feasibility region more significantly than k_{iP2}. Comparing to Fig. 4.5c, where the feasibility region is significantly (and visibly) reduced when $k_{pV_{ac}2}$ is increased by 50% (from 0.2 to 0.3 p.u.), there is negligible change in the feasibility region when k_{iP2} also increases by the same relative amount (50% from 120 to 180 p.u.) as shown in Fig. 4.5d.

In general, the results clearly show that the parameters with high absolute correlation index values have a significant impact on the system's feasible region boundary and are considered as influential parameters.

Table 4.3 Partial Spearman correlation coefficient for each pair of bifurcation parameters. © 2023 IEEE, reprinted, with permission, from Chen et al. [3]

	ω_{pll2}	L_{12}	Z_{13}	k_{pP2}	$k_{pV_{ac}2}$	k_{iP2}	$k_{iV_{ac}2}$
ω_{pll2}	1.00	– 0.51	– 0.71	0.78	– 0.80	– 0.05	– 0.16
L_{12}	– 0.51	1.00	– 0.45	0.50	– 0.48	– 0.01	– 0.08
Z_{13}	– 0.71	– 0.45	1.00	0.70	– 0.68	– 0.03	– 0.09
k_{pP2}	0.78	0.50	0.70	1.00	0.75	0.15	0.24
$k_{pV_{ac}2}$	– 0.80	– 0.48	– 0.68	0.75	1.00	– 0.02	– 0.14
k_{iP2}	– 0.05	– 0.01	– 0.03	0.15	– 0.02	1.00	0.09
$k_{iV_{ac}2}$	– 0.16	– 0.08	– 0.09	0.24	– 0.14	0.09	1.00

4.3.2.4 Boundaries in the $k_{pP2} - k_{pV_{ac}2}$ Plane

In addition to the common negative correlations between bifurcation parameter pairs, Table 4.3 shows that system parameters may have a positive index. Table 4.3 indicates that the partial Spearman correlation coefficient between $k_{pP2} - k_{pV_{ac}2}$ is 0.75. This value indicates that the relationship between $k_{pP2} - k_{pV_{ac}2}$ is positive. Therefore, if both k_{pP2} and $k_{pV_{ac}2}$ increase, stability margins should also increase. The calculated partial Spearman index also reveals potential interactions between the power control and voltage control in VSC_2. If the outer control loops of VSC_2 are entirely decoupled, the stability limit with $k_{pV_{ac}2}$ should not be influenced by any changes in k_{pP2} and, therefore, they should not have any interdependency. However, the obtained correlation coefficient value is far from zero, which suggests some interaction among the outer control loops of VSC_2.

The feasibility region formed in the $k_{pP2} - k_{pV_{ac}2}$ plane is shown in Fig. 4.8, considering (a) different values of coupling inductances, L_{12}, and (b) different values of grid impedances, Z_{13}. With all other parameters fixed, the parameter spaces that guarantee system stability are the areas beneath the curves.

From the results shown in Fig. 4.8a, the system moves away from instability with increases in k_{pP2} and additional regulation space is left for k_{pP2}. This verifies the positive partial Spearman correlation index between k_{pP2} and $k_{pV_{ac}2}$. Moreover, Fig. 4.8 shows that the system's stable operational area is reduced with a higher value of Z_{13} or L_{12}.

4.3.3 *Parameters in Two VSCs*

This section considers the control parameters in both VSCs and network parameters as bifurcation parameters. The feasibility region is first established by conducting a similar analysis process as in the case of considering the parameters of a single VSC (as discussed in Sect. 4.3.2). However, by considering variations in the parameters

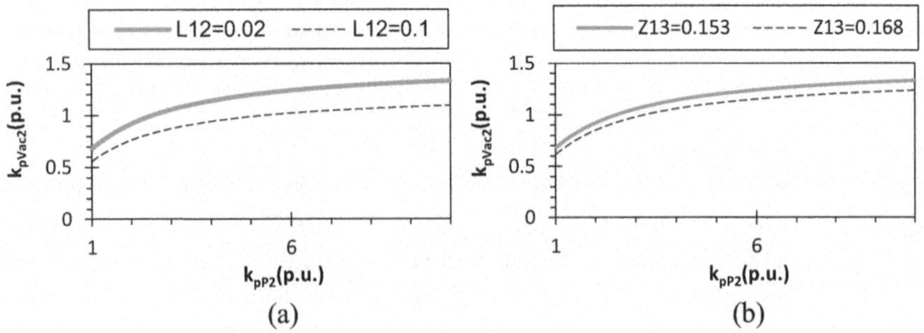

Fig. 4.8 Feasibility region boundaries on the 2D plane of $k_{pP2} - k_{pV_{ac}2}$ with **a** variation in L_{12} and **b** variation in Z_{13}. © 2023 IEEE, reprinted, with permission, from Chen et al. [3]

of both VSCs, it is possible to identify interactions between the VSCs. The partial Spearman indices of system parameters are then calculated for all parameter pairs to select those showing strong correlation (and high interaction). In a two-VSC case, the following twelve parameters are considered:

1. The bandwidth of the PLL in both VSCs: ω_{pll1}, ω_{pll2}
2. The grid impedance: Z_{13}
3. The coupling inductance: L_{12}
4. Both VSC power control proportional gains: k_{pP1}, k_{pP2}
5. Both VSC voltage control proportional gains:$k_{pV_{ac}1}$, $k_{pV_{ac}2}$
6. Both VSC power control integral gains: k_{iP1}, k_{iP2}
7. Both VSC voltage control integral gains: $k_{iV_{ac}1}$, $k_{iV_{ac}2}$.

4.3.3.1 Boundaries in the $\omega_{pll1} - \omega_{pll2}$ Plane

The details of the correlation coefficients considering variations in the parameters of two VSCs are included in Appendix B.1. The calculated partial Spearman correlation index suggests a negative relationship between ω_{pll1} and ω_{pll2} (with a value of – 0.61). This also further implies interaction between the two VSCs, specifically through the PLL bandwidth.

The feasibility regions formed by $\omega_{pll1} - \omega_{pll2}$ with different parameter settings are shown in Fig. 4.9. From the charts, the area below the curve is the system feasibility region, with all other parameters at nominal values at the lower bound of the search range. As is shown in Fig. 4.9, when the PLL bandwidth of one VSC is increased, the PLL of another VSC should reduce its bandwidth to ensure system stability. This is a further confirmation of the negative correlation between ω_{pll1} and ω_{pll2}. These diagrams also illustrate that variations in other bifurcation parameters will affect the system feasibility region. In particular, when the grid impedance (Z_{13}) and the integral gain of the power control loops of both VSCs (k_{iP}) are increased, the system's stable region is reduced considerably.

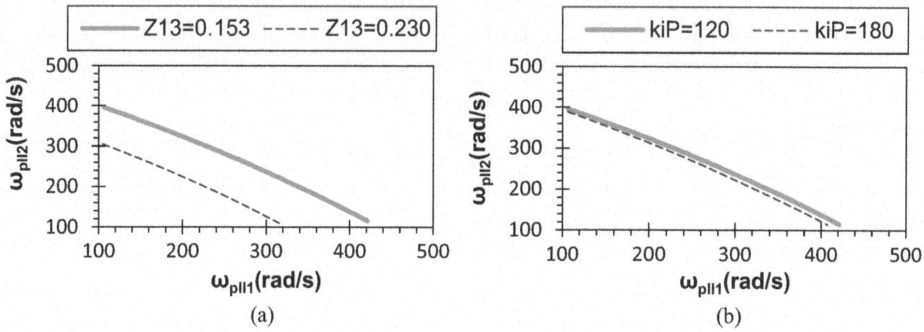

Fig. 4.9 Feasibility region boundaries on the 2D plane of $\omega_{pll1} - \omega_{pll2}$ with **a** variation in Z_{13} and **b** variation in k_{iP}. © 2023 IEEE, reprinted, with permission, from Chen et al. [3]

As discussed in Sect. 4.3.2.3, a larger absolute partial Spearman correlation coefficient indicates a stronger relationship between parameter pairs and further signifies that the key parameters will have significant impacts on the system boundary. The calculated correlation indices between $\omega_{pll1} - Z_{13}$ (−0.65) and $\omega_{pll2} - Z_{13}$ (−0.75) are much larger than the coefficients between $\omega_{pll1} - k_{iP}$ (−0.04) and $\omega_{pll2} - k_{iP}$(−0.08). This is illustrated by the plots shown in Fig. 4.9a, b. In Fig. 4.9a, the system feasibility region below the feasibility region boundary is reduced significantly when Z_{13} increases by 50% (from 0.153 to 0.230 p.u.), while Fig. 4.9b shows that the system boundary is hardly changed when k_{iP} increases by the same 50% margin (from 120 to 180 p.u.). Hence, compared with k_{iP}, Z_{13} is considered a key parameter that could significantly affect the system boundary. A similar approach can be applied to identify $k_{pV_{ac}}$ as a key parameter.

Hence, the results again demonstrate that large absolute partial Spearman correlation coefficient values reveal the critical parameters affecting the system's feasiblility region boundary.

4.3.4 Time-Domain Simulation

This section verifies the previous analytical system feasibility region boundary results by applying time-domain simulations. The system feasibility areas, formed by k_{pP2} and $k_{pV_{ac}2}$ with different values of coupling inductance L_{12}, are shown in Fig. 4.10.

The solid curve represents the feasibility region boundary when $L_{12} = 0.02$ p.u., and the dashed curve illustrates the feasibility region boundary when $L_{12} = 0.1$ p.u. The system feasibility region is the area below the corresponding stability curves. Three system operating points (i.e. A (5.0, 0.8), B (5.0, 1.1), and C (5.0, 1.3)) are selected in Fig. 4.10. Based on the identified feasibility region boundaries, when the system is operating at point A, it will be stable for both the case where $L_{12} = 0.02$ p.u. and the case where $L_{12} = 0.1$ p.u. When the system operating point moves to point

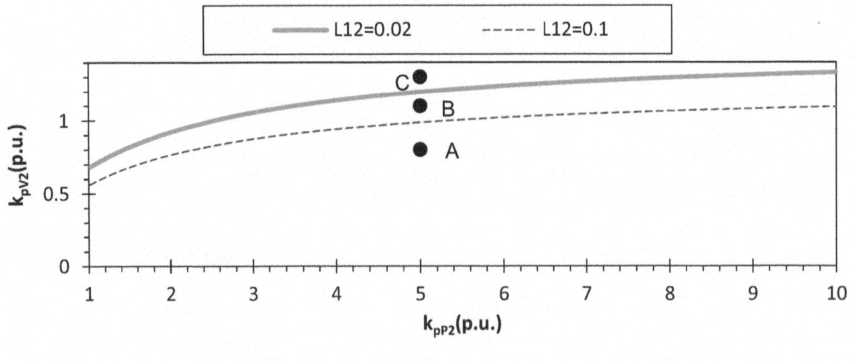

Fig. 4.10 Selection of three operating points on the $k_{pP2} - k_{pV_{ac2}}$ plane for verification of time-domain simulation. © 2023 IEEE, reprinted, with permission, from Chen et al. [3]

C, the power system is unstable in both cases, as point C is located outside of both feasibility regions. For point B, analytical results show that the system will be stable when $L_{12} = 0.02$ p.u. (below the solid line) and unstable when $L_{12} = 0.1$ p.u. . (above the dashed line).

The time-domain simulation results of the six scenarios (the points A, B, and C with $L_{12} = 0.02$ p.u. and $L_{12} = 0.1$ p.u.) are illustrated in Fig. 4.11. A small disturbance, in which the power references of the VSCs decrease from 1.0 to 0.9 p.u., is injected into the system at 0.05 s. The vertical axis of the charts is the power output of the VSCs, and the horizontal axis represents time in seconds. For point A, with $L_{12} = 0.02$ p.u. and $L_{12} = 0.1$ p.u., the power output of the VSCs will quickly stabilise at 0.9 p.u. after the set point change, as shown in Fig. 4.11a, b. If the system operation point moves to point B, Fig. 4.11c illustrates that the system is stable again after the small disturbance if $L_{12} = 0.02$ p.u. However, if L_{12} increases to 0.1 p.u., the system will become unstable after the small disturbance (Fig. 4.11d). Finally, for point C, as illustrated in Fig. 4.11f, the system oscillates with increasing magnitude and finally becomes unstable with $L_{12} = 0.02$ p.u. and $L_{12} = 0.1$ p.u. These results verify the analytical results discussed previously, demonstrating that the proposed methodology can efficiently identify the N-dimensional system feasibility region by taking N different parameters into consideration. Additionally, it also indicates that the interpolated hypersurface can be used to generate the feasibility boundary accurately, as the analysis based on results produced using the interpolated hypersurface is consistent with the relevant nonlinear time-domain simulations.

4.4 Feasibility Analysis of a Large System

The initial studies performed using the small-sized test system discussed in the last section have clearly demonstrated the effectiveness of the proposed methodology. A brief illustrative example, a modified IEEE 39-bus system containing two VSCs, is presented in this section to further demonstrate the applicability of the developed analysis framework to a larger system. The same evaluation methodology used when studying the small-sized test systems is adopted again to establish the feasibility region in large systems. The partial Spearman correlation coefficients of all parameter pairs can again be used to identify critical system parameters with strong correlations. For this larger test system, the following fourteen parameters are considered:

1. The bandwidth of the PLL in both VSCs: ω_{pll1}, ω_{pll2}
2. Both VSC power control proportional gains: k_{pP1}, k_{pP2}
3. Both VSC voltage control proportional gains: $k_{pV_{ac}1}$, $k_{pV_{ac}2}$
4. Both VSC power control integral gains: k_{iP1}, k_{iP2}
5. Both VSC voltage control integral gains: $k_{iV_{ac}1}$, $k_{iV_{ac}2}$
6. VSC adjacent transmission line impedance: Z_{1-2}, Z_{2-3}, Z_{2-25}, Z_{25-26}.

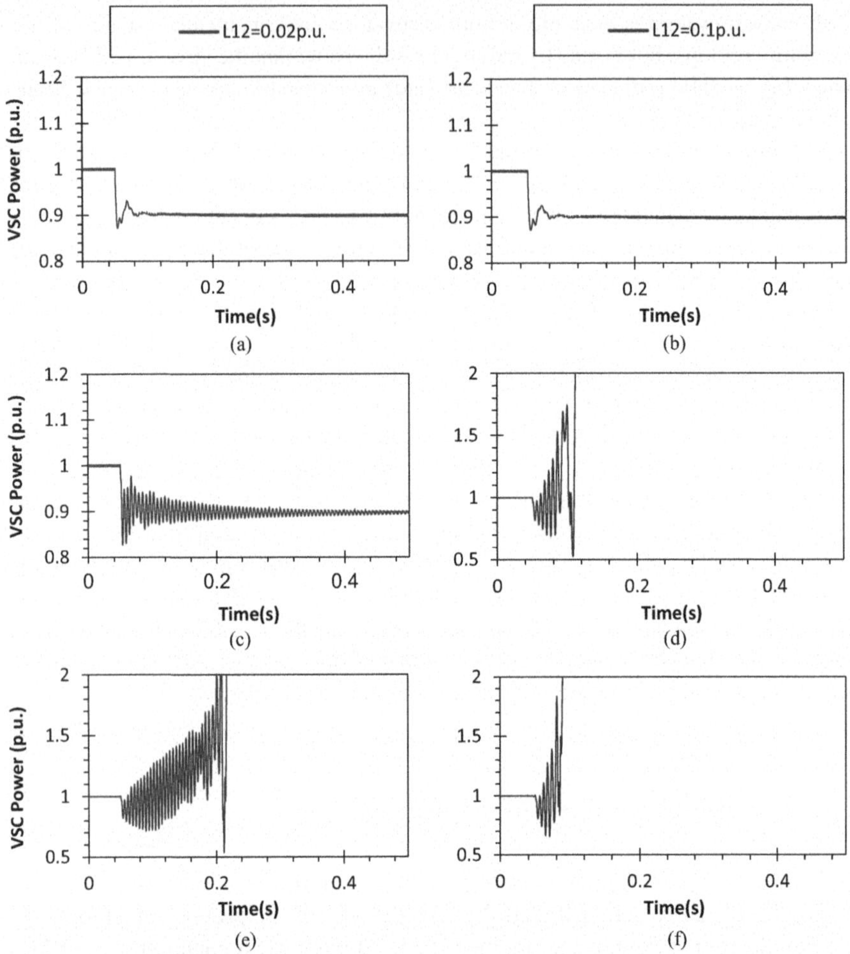

Fig. 4.11 Verification of time-domain simulation with **a** point A when $L_{12} = 0.02$ p.u., **b** point A when $L_{12} = 0.1$ p.u., **c** point B when $L_{12} = 0.02$ p.u., **d** point B when $L_{12} = 0.1$ p.u., **e** point C when $L_{12} = 0.02$ p.u., and **f** point C when $L_{12} = 0.1$ p.u. © 2023 IEEE, reprinted, with permission, from Chen et al. [3]

4.4.1 Test Network Details

Two VSCs are connected to Buses 2 and 25 to replace G30 and G37. The VSCs have the same capacity as the original SGs. The modified network is shown in Fig. 4.12, with VSC parameters given in Appendix A.4.

To investigate the system's small-disturbance stability, the transmission lines shown in Fig. 4.12 are modelled with ODEs to create a dynamic network model that includes the network LC dynamics, hence capturing potential interactions between the network and the VSCs.

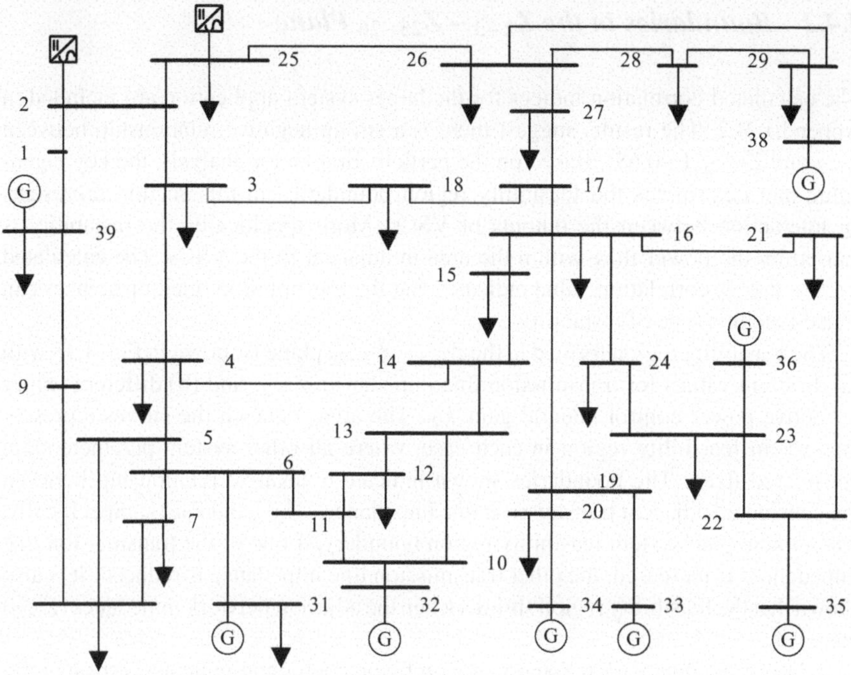

Fig. 4.12 Large-scale test system with multiple VSCs connected

Similar to Tables 4.2 and 4.4 summarises the search range of fourteen bifurcation parameters considered within the large system. Again, the initial values of the controller bifurcation parameters shown in Table 4.4 are tuned based on the SIMC. The same search range adopted in the small-sized test system (between the initial value and ten times the initial value) is used to analyse the multi-dimensional feasibility region boundary.

Table 4.4 Key system parameter search ranges for the large-size test network

Key system parameter	System parameter search range
$\omega_{pll(1,2)}$ (rad/s)	[100, 1000]
$k_{pP(1,2)}$ (p.u.)	[1, 10]
$k_{pV_{ac}(1,2)}$(p.u.)	[0.2, 2]
$k_{iP(1,2)}$ (p.u.)	[120, 1200]
$k_{iV_{ac}(1,2)}$ (p.u.)	[215, 2150]
Z_{1-2} (p.u.)	[0.0412, 0.142]
Z_{2-3} (p.u.)	[0.0152, 0.152]
Z_{2-25} (p.u.)	[0.0111, 0.111]
Z_{25-26} (p.u.)	[0.0325, 0.325]

4.4.2 Boundaries in the $Z_{2-3} - Z_{25-26}$ Plane

The calculated correlation indices for the larger system application are included in Appendix B.2. The results suggest there is a strong negative relationship between Z_{2-3} and Z_{25-26} (−0.68). Based on the participation factor analysis, the key eigenvalue that determines the feasibility region boundaries in this section represents an interaction between the outputs of VSCs. More specifically, the instability is caused by the power flow within the area in adjacent to the VSCs. The calculated $Z_{2-3} - Z_{25-26}$ correlation value indicates that the transmission line impedances can exacerbate this type of instability.

The feasibility region formed in the $Z_{2-3} - Z_{25-26}$ plane is shown in Fig. 4.13 with (a) different values for transmission line impedance, Z_{1-2}, and (b) different values for active power control integral gain, k_{iP}. The areas beneath the curves represent the system feasibility region in each case, where all other system parameters are considered fixed. The boundaries shown indicate a negative relationship between the considered adjacent transmission line impedances, Z_{2-3} and Z_{25-26}. Specifically, when tracing the system feasibility region boundary, if one of the transmission line impedances is increased, the other transmission line impedance is reduced. It is also shown that the feasibility region shrinks when the adjacent network impedance (Z_{1-2}) increases.

A larger absolute partial Spearman correlation coefficient indicates a strong relationship between parameters, which has a significant impact on the system feasibility region. The correlation between Z_{1-2}–Z_{2-3} (−0.62) and Z_{1-2}–Z_{25-26} (−0.79) far exceeds that between $k_{iP} - Z_{2-3}$ (0.06) and $k_{iP} - Z_{25-26}$ (0.07). This is verified in Fig. 4.13a, b. In Fig. 4.13 (a), the system feasibility region is visibly reduced with a 50% increase in Z_{1-2}. However, in Fig. 4.13b, there is negligible change in the system feasibility boundary when the integral gains in the voltage control of both VSCs increase by 50%. Hence, unlike k_{iP}, Z_{1-2} is a critical parameter and has a significant impact on the system feasibility region.

In general, for the range of parameters considered, the calculated partial Spearman correlation indices reveal that the impedances of the transmission lines adjacent to

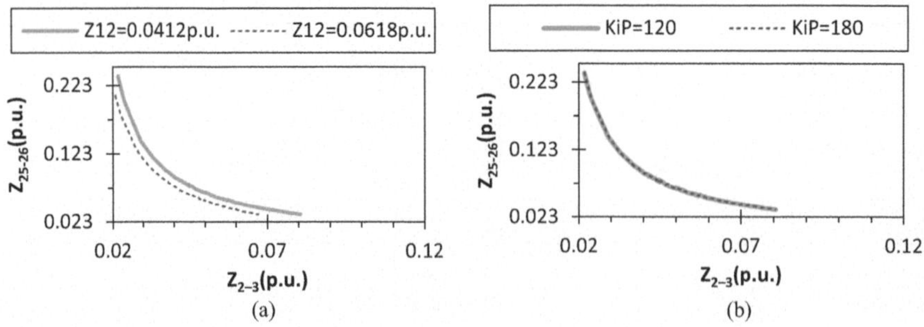

Fig. 4.13 Feasibility region boundaries on the 2D plane of $Z_{2-3} - Z_{25-26}$ with **a** variation in Z_{1-2} and **b** variation in k_{iP}. © 2023 IEEE, reprinted, with permission, from Chen et al. [3]

the VSCs have a more significant impact on system feasibility regions than the VSC controller parameters. This aligns with the results from the application within a small-size system, where network impedance was found to have the largest influence on the system feasibility region boundary.

Previous research in [10] indicates that interactions among VSCs become significant under weak grid conditions where the short-circuit ratio (SCR) is low. In the large system, the VSCs are operating in a stable zone, and the eigenvalues are far away from the imaginary axis. The SCR in the VSC connection area is high. Hence, variations in the VSC controller parameters have negligible impacts on the system feasibility region in the large system, and the resulting interactions between VSCs are much less significant. Further simulations and analysis, which will be performed in detail in Chap. 5 will reveal that under low-SCR conditions, the interactions between VSCs become more significant, and that the control parameters begin to have more obvious impacts on the system feasibility region boundary.

4.5 Summary

An analysis framework that can identify and analyse multi-dimensional feasibility region boundaries in a power system with multiple VSCs has been established and tested in this chapter. The proposed framework consists of two parts: (1) identifying the system critical points considering N different parameters varying simultaneously using the continuation method and (2) analysing the obtained data points using a partial Spearman correlation coefficient. The analysis mainly focuses on the small-disturbance stability of the power system. With the developed analysis framework, power system operators can efficiently identify the system's feasible boundary in N-dimensional space. The analytical results provide vital information for power system operators to select parameter configurations that guarantee stable system operation.

Rather than directly presenting the system feasibility boundary using a generalised mathematical equation, the developed methodology adopts a statistical approach to identify the multi-dimensional feasibility boundary. Based on the continuation method and the Halton sequence, the developed methodology continuously calculates the system stability limit points, which are located on the system feasibility boundary hypersurface. Those points are then used to form an interpolated hypersurface using interpolation techniques to approximate the real system feasibility boundary in N-domains. An illustrative example shows that the interpolated system feasibility boundary hypersurface can be used to produce results as accurate as those obtained by using the continuation method. The relative root-mean-square deviation of the curve derived from the interpolated hypersurface is just 5.56×10^{-3}, which is smaller than the acceptable relative tolerance (10^{-2}) established in this chapter. This clearly demonstrates the effectiveness of the proposed methodology in identifying the system's multi-dimensional feasible region boundary. Hence, once the interpolated multi-dimensional feasibility boundary is completed, it will be used for further sensitivity analysis.

The partial Spearman correlation coefficient, which offers a global sensitivity analysis technique, is used to quantify the relationships between bifurcation parameter pairs and to numerically identify influential parameters based on the obtained data points. This analysis technique accounts for sensitivity over the full parameter range, as well as the interdependency among parameters, allowing a comprehensive assessment of the multi-dimensional feasibility region boundary of the system.

A detailed study of the system feasibility region using the proposed analysis framework is presented on a small-sized test system with the consideration of parameter variations in (1) one VSC and (2) two VSCs. It has been clearly demonstrated that key parameters such as the VSC adjacent impedance and the bandwidth of the PLL have large correlation values. Therefore, those parameters are considered as the influential parameters affecting the system's N-dimensional feasibility boundary significantly. Additionally, the bandwidth of PLLs from different VSCs has a high absolute correlation coefficient value of − 0.61. This indicates the interactions among the control loops of the VSCs. Apart from that, the sign of the correlation represents the relationship between parameters in each case. For example, the negative sign in the $\omega_{pll2} - Z_{13}$ correlation suggests that if ω_{pll2} increases, Z_{13} is likely to decrease at the feasibility region boundary. Aside from such conventional negative correlations, there also exist positive correlations among some parameters, such as $k_{pP2} - k_{pV_{ac}2}$. In the positive correlation case, increasing the value of parameters leads to an increase in the system stability margin. A brief study on a modified IEEE 39-bus system has shown that the impedance of the transmission lines around the VSCs has a significant impact on system stability, with other control parameters having only negligible influence on dynamic system behaviour. However, the sensitivity analysis results appear to depend on the grid strength, particularly when a large power system is considered, which requires further in-depth study. A further sensitivity assessment of system stability limit in large power systems considering different grid conditions will be performed in the next chapter.

This chapter mainly focuses on small-disturbance stability analysis of power systems based on bifurcation theory. Combined with the studies discussed in Chap. 3, this chapter clearly demonstrates the capability of bifurcation theory-based approaches in analysing system small-disturbance and voltage stability, with each type of instability typically analysed individually, as discussed in these chapters. However, newly emergent power electronics-related issues have blurred the boundary between different types of stability, leading to a requirement to consider different types of stability simultaneously when performing system stability assessments. This can be achieved by the application of bifurcation theory in which different instabilities can be presented by various types of bifurcations. The system stability study in the next chapter will thus consider multiple instability mechanisms concurrently with the application of bifurcation theory.

References

1. Zhou JZ, Ding H, Fan S, Zhang Y, Gole AM (2014) Impact of short-circuit ratio and phase-locked-loop parameters on the small-signal behavior of a VSC-HVDC converter. IEEE Trans Power Deliv 29(5):2287–2296
2. Harnefors L, Bongiorno M, Lundberg S (2007) Input-admittance calculation and shaping for controlled voltage-source converters. IEEE Trans Ind Electron 54(6):3323–3334
3. Chen Y, Preece R, Barnes M (2023) Identifying feasibility region boundaries in power systems with multiple VSCs. IEEE Trans Power Syst 38(2):1229–1241
4. Guckenheimer J, Holmes P (1983) Nonlinear oscillations, dynamical systems, and bifurcations of vector fields, 1st edn. Springer, New York, NY
5. Venkatasubramanian V, Schättler H, Zaborszky J (1995) Dynamics of large constrained nonlinear systems-A taxonomy theory. Proc IEEE 83(11):1530–1561
6. Venkatasubramanian V, Schättler H, Zaborszky J (1995) Local bifurcations and feasibility regions in differential-algebraic systems. IEEE Trans Automat Contr 40(12):1992–2013
7. James E (2003) Gentle, random number generation and monte carlo methods, 2nd edn. Springer, New York, NY
8. Wang X, Hickernell FJ (2000) Randomized Halton sequences. Math Comput Model 32(7–8):887–899
9. Kocis L, Whiten WJ (1997) Computational investigations of low-discrepancy sequences. ACM Trans Math Softw 23(2):266–294
10. Bhat CR (2003) Simulation estimation of mixed discrete choice models using randomized and scrambled Halton sequences. Transp Res Part B Methodol 37(9):837–855
11. Amidror (2002) Scattered data interpolation methods for electronic imaging systems: a survey. J Electron Imaging 11(2):157–176
12. Steffensen JF (2006) Interpolation, 2nd edn. Dover Publications, New York
13. Kurowicka D, Cooke R (2006) Uncertainty analysis with high dimensional dependence modelling, 1st ed. John Wiley & Sons, Ltd. Chichester, UK
14. Myers JL, Well AD, Lorch Jr RF (2010) Research design and statistical analysis, 3rd ed. Lawrence Erlbaum Associates, Mahwah, N.J.
15. Iooss B, Lemaître P (2015) A review on global sensitivity analysis methods. In: Uncertainty management in simulation-optimization of complex systems: algorithms and applications, vol 59, Springer New York LLC, pp 101–122
16. Fisher RA (1924) The distribution of the partial correlation coefficient. Metron 3:329–332
17. Wan C, Huang M, Tse CK, Ruan X (2015) Effects of interaction of power converters coupled via power grid: A design-oriented study. IEEE Trans Power Electron 30(7):3589–3600
18. Huang Y, Wang D, Shang L, Zhu G, Tang H, Li Y (2017) Modeling and stability analysis of DC-link voltage control in multi-VSCs with integrated to weak grid. IEEE Trans Energy Convers 32(3):1127–1138
19. Skogestad S (2003) imple analytic rules for model reduction and PID controller tuning. J Process Cont 13(4):291–309

Chapter 5
Identification of Critical Parameters Affecting System Loadability

Chapters 3 and 4 clearly demonstrated the capability of bifurcation theory in analysing the small-disturbance stability and voltage stability of a system. Although much of the research literature outlined in previous chapters has investigated multiple types of system stability, it typically analyses each classification of stability individually [1, 2]. A large number of converters connected to future power systems means that potential interaction between these power electronics components and the AC network may result in numerous issues [3]. These potential interactions have driven voltage stability oscillations to a higher frequency range and have further blurred the boundary between voltage and small-disturbance stability. This implies that the different types of stability become interdependent and interact with each other. The work in [4, 5] suggests that a system may encounter small-disturbance stability issues before reaching the voltage limit when considering the system loadability, particularly when multiple VSCs are introduced into the system. This was shown to be true in the case of the power system blackout in Pakistan in 2006, when small-disturbance instability and voltage instability occurred simultaneously [6]. Therefore, it is becoming increasingly important to consider both voltage stability and small-disturbance stability when analysing future power systems.

There has been a great deal of interest in the identification of critical system parameters based on their influence on system stability limits, which could help system operators to optimise the dynamic performance of a system using minimum computational resources. A review of the application of sensitivity analysis tools in power systems is presented in Sect. 1.5.3. Although various global sensitivity analysis techniques have been adopted to identify influential parameters, system parameter variations are currently limited to a small operating range, particularly around one operating point. Taking the work in [1] as an example, investigations are typically performed with a relatively constant grid strength. Such results should be considered local, as the impact of various grid strengths is ignored. As is discussed in Sect. 4.4.2, the grid strength at which power electronics are connected may have a significant impact on the sensitivity of analysis results, particularly for large-scale systems with

Y. Chen, *Stability Assessment of Power Systems with Multiple Voltage Source Converters*, Springer Theses, https://doi.org/10.1007/978-3-031-63095-8_5

multiple VSCs. Hence, there is a need to identify influential system parameters over a wide range by considering the influence of different grid strengths.

Previous studies have investigated the impact of different VSC controllers on the system stability considering both strong and weak grid conditions [7, 8]. However, most of the analysis is performed around one specific operating point under each grid condition. Occasionally, a limited number of scenarios have been considered, but no thorough assessment of VSC control over more wide-ranging grid conditions has yet been completed. This would require a thorough assessment of the performance of different VSC control schemes under various scenarios to establish how the VSC controls affect the loading margin.

This chapter uses bifurcation theory to allow the joint consideration of the small-signal stability, as well as voltage stability, when assessing the system loadability, which represents the fourth contribution of this thesis. Within bifurcation theory, small-disturbance and voltage instabilities can be identified as different types of bifurcation points along the trajectory of system operation. This represents an update to the formulation of system loadability limits by considering different types of stability mechanisms simultaneously using bifurcation theory.

A similar statistical assessment framework to that adopted in Chap. 4 (when small-disturbance stability is considered) is used in this chapter to identify the most influential parameters based on their impact on system loadability. The numerical method involves quasi-Monte–Carlo simulations of a wide range of grid scenarios that consider uncertainties in system parameters, grid strengths and VSC control schemes. Within each grid scenario, the system loadability trajectory is computed using the continuation method, and a standard loadability assessment has also been undertaken. The numerical results obtained through this method are analysed using partial Spearman correlation (discussed in Chap. 4) to identify the parameters influencing the system loadability.

The developed framework can also be used to evaluate the performance of VSC control schemes across wide-ranging scenarios, which represents the fifth contribution of this thesis. Through this evaluation, this chapter establishes the VSC control scheme that provides the largest loading margin under various scenarios. The evaluations of VSC control schemes are performed on a modified IEEE 39-bus system in which single and multiple VSCs are considered.

5.1 Proposed Methodology

This section introduces a methodology that can be used either to identify the influential parameters affecting system loadability or to perform a thorough assessment of VSC controllers over a broad range of system conditions. Partial Spearman correlation coefficients are used to numerically rank parameters based on their impact on the system loadability across wide-ranging loadability scenarios. This methodology can also be used to provide an evaluation of the performance of the VSC controllers over a diverse range of loadability cases, thereby determining the control scheme for system operation to provide the widest feasible range.

5.1.1 Establishing the Initial Grid Conditions

To begin with, the uncertainties in system parameters, including transmission line impedance and the control parameters of VSCs, are used to generate a set of network initial scenarios. The details of the parameters considered in this chapter are as follows:

1. The impedance of the transmission line: $Z_{i,j}$.
2. The bandwidth of the VSC PLL: ω_{plli}.
3. The proportional gain of the VSC power control loop: k_{pPi}.
4. The proportional gain of the VSC reactive power/voltage control loop: $k_{pQi}/k_{pV_{ac}i}$.
5. The integral gain of the VSC power control loop: k_{iPi}.
6. The integral gain of the VSC reactive power/voltage control loop: $k_{iQi}/k_{iV_{ac}i}$.

The subscript i in the notation represents the parameter of the ith VSC.

In this study, the variations of all system parameters are modelled using a uniform distribution. The value range of each parameter considered will be discussed in the following section. It should be noted that in this chapter, no variation in the initial loading values (of all loads) is considered, as their impact on the system loadability is significant. Given such a large correlation index value, the relationships between other system parameters and the system loadability are unrepresented. Specifically, the strong correlation between the load and the system loadability covers the impact of the other parameters on system loadability. Therefore, no variation in the initial loading of the system is considered in this chapter.

In order to comprehensively investigate the full range of potential initial conditions, it is necessary to consider variations in all system parameters. For a sensitivity study, where the underlying probability distribution of the variation is not considered, a uniform distribution across the range of possible values for each parameter is appropriate. One approach is to divide the potential range of each parameter into a number of representative values, and then exhaustively sample every combination of parameter values. However, with 51 parameters considered in this section, this is not practical due to the extremely high computational requirements required. As demonstrated in Chap. 4, a quasi-random search can fully explore the parameter space, and hence, Halton sequences are used again to efficiently sample the space. These sampling points provide a representative series of scenarios that describe the initial network conditions.

5.1.1.1 Value Range of Parameter

Within this section, the transmission line impedance varies between the initial value and 1.2 times the initial value, and the control parameter of VSCs varies between the initial value and 10 times the initial value. Based on the studies undertaken in this research, an expansion of the transmission line impedance range to twice the initial value or higher significantly increases the likelihood of an unstable system loadability curve, and no feasible operating points are available. Such scenarios are of little value for analysis and would be of less interest to system operators, as no stable operating point exists on the loadability trajectory of the system. Therefore, any further extension of the search range of transmission line impedance would reduce the number of scenarios available for analysis. A similar explanation can be applied to justify the selection of the search range of VSC control parameters. In general, the parameter search range has been carefully selected to encompass a wide range of grid initial scenarios possible that can yield feasible operating points, though some of the network initial conditions are too extreme to be found in real power systems.

5.1.1.2 Number of Simulations

An important factor affecting the accuracy of such quasi-Monte–Carlo simulations is the number of simulations completed. The results in [9] show that the correlation assessment becomes increasingly more accurate as the number of simulations increases, but this comes at the cost of an increasingly heavy computational burden. With the consideration of the simulation time and the recommendations from [9], each test case correlation assessment contains 10^4 scenarios.

5.1.2 Computing the System Loadability

Following each initial grid condition, the system loadability limit is computed using the continuation method developed in Chap. 2. The continuation method is selected for this study because it provides the most detailed information about the system trajectory in terms of multiple instability mechanisms.

5.1.3 Identification of Influential Parameters

The partial Spearman correlation coefficient is calculated to show the correlation between the system loadability and the system parameters. The calculated correlation indices quantify the impact of the system parameters on loadability. Parameters correlating strongly with the feasible range are deemed key parameters affecting system loadability.

5.1.4 *Performance Assessment of Different VSC Control Schemes*

The performance of different controllers under each scenario can also be assessed based on loadability analysis to study their performance across a wide range of loadability conditions. Specifically, the study will establish the controller that offers the widest available operating range under various loadability scenarios.

5.1.5 *Detailed Steps*

The detailed steps for applying the proposed approach are as follows:

Step 1: Generate various initial network scenarios based on the quasi-random sampling of the stochastic sources of system uncertainties shown in Sect. 5.1.1.

Step 2: Compute the system loadability curve using the continuation method for each initial operating condition and assess the resulting loadability trajectory based on bifurcation theory. This will generate abundant loadability data considering wide-ranging grid conditions.

Step 3: (a) The sensitivity analysis tool, the partial Spearman correlation coefficient, is employed to identify key parameters which have significant influences on system loadability.

(b) Alternatively, the performance of VSC control schemes can be assessed based on their impact on the system loadability under the same initial grid conditions.

5.2 Application and Analysis of a System Containing a Single VSC

The previous sections have outlined a methodology to identify critical parameters based on their influence on the system's loadability, and to perform assessments for VSC controllers. This section applies the method to a test network with a single VSC connected. Firstly, the method is used to identify influential parameters considering different VSC control schemes (PQ and PV_{ac}) and grid strengths. The performance of different VSC controllers (i.e., PQ and PV_{ac}) is then evaluated over a wide range of loadability conditions.

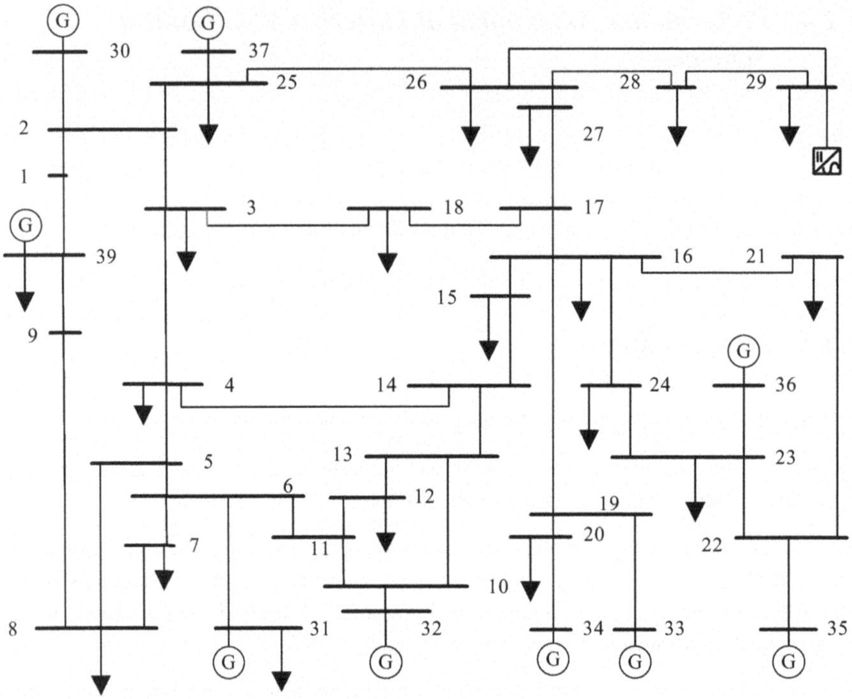

Fig. 5.1 The IEEE 39-bus system modified for one VSC

5.2.1 *Modification of the Test Network*

In this section, a study is performed using a modified IEEE 39-bus system with a single VSC connected. The modified test network, as shown in Fig. 5.1, features a VSC connected to Bus 29 instead of the original SG integrated at Bus 38. Data and controller settings for the VSC are provided in Appendix A. The test network in this section adopts dynamic network modelling (developed in Sect. 2.1.7), in which the entire network is modelled with ODEs to capture higher frequency oscillations resulting from interactions between the VSC and the network.

5.2.2 *Identification of Influential Parameters*

A typical loadability curve is shown in Fig. 5.2. Note that the axes are dimensionless, since Fig. 5.2 represents the general shape of the typical system equilibrium curve of interest in this study rather than a specific example. Along the system loadability trajectory, three types of bifurcations are shown: an SNB and two HBs. The SNB represents the maximum loading point, beyond which the system equilibrium point

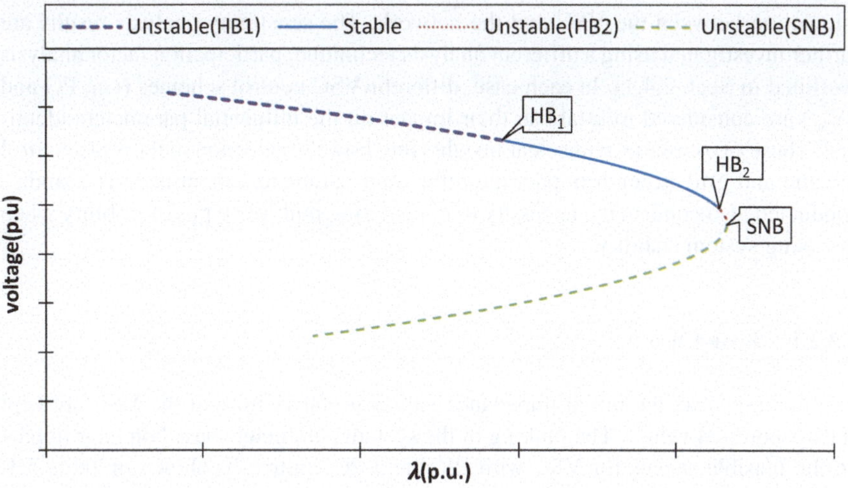

Fig. 5.2 A typical system loadability curve

disappears. The two HBs represent two typical eigenvalues. HB$_1$ represents an eigenvalue that is unstable at the initial point but becomes stable as the loading increases. HB$_2$ represents a mode that is initially stable but becomes unstable as the loading increases.

The feasible range over which the system can operate is from HB$_1$ to HB$_2$ in this typical system equilibrium trajectory example. Therefore, the locations of the different bifurcations, HB$_1$, HB$_2$, and SNB, jointly determine system loadability. Consequently, the identification of system critical parameters is to some degree equivalent to determining the influential parameters affecting the locations of these bifurcations. It should be noted that loadability curves in all scenarios contain an SNB, but not all scenarios consist of two types of HB along the system loadability trajectories simultaneously. Some scenarios may include only one type of HB or even no HBs on their loadability curves.

The partial Spearman correlation coefficient will be computed to show the interdependency between each parameter and the feasible range. The calculated correlation indices provide a quantitative measurement of the impact of parameters on the system loadability. A high absolute correlation value indicates that variations in the parameters are more likely to result in changes in the feasible range, which further influences system loadability. Hence, the parameters that have a high absolute correlation index are considered key parameters. System parameters will be ranked based on the calculated absolute partial Spearman correlation value in the following sections.

This section considers two cases: (i) a base case under nominal operating conditions and (ii) a weak-grid case in which the strength of the system is reduced. In the latter case, the impedance of the transmission lines in the vicinity of the VSC is increased to reduce the SCR seen by the VSC and therefore to exacerbate the interaction between the VSC and the network. The sensitivity analysis results are further investigated using a different analysis technique, participation factor analysis (outlined in Sect. 2.4.1). In each case, different VSC control schemes (i.e., PQ and PV_{ac}) are considered to establish their impact on the influential parameters identified. These two cases provide clear insights into how the selection of the VSC control scheme and grid strength impact the influential parameters identified. The studies conducted also confirm the necessity of considering multiple types of stability when assessing system stability.

5.2.2.1 Base Case

For the base case, the initial impedance values in the vicinity of the VSC are kept at their nominal values. The ranking of the system parameters based on their impact on the feasible ranges for VSC with PV_{ac} and PQ controls is shown in Table 5.1. For brevity, only the top five parameters affecting the feasible ranges are listed in the table, alongside their corresponding correlation index values. The full ranking of state variables based on their influence on the feasible range is provided in Appendix B.3.

When the VSC operates with PV_{ac}, Table 5.1 shows that the impedances of transmission lines Z_{6-31}, Z_{10-32}, and Z_{3-4} are strongly correlated with the feasible operating range, with values of – 1.00, – 0.95, and – 0.80, respectively.

For the VSC with PV_{ac} in the base case, both HB_1 and HB_2 disappear, and only the SNB exists in all loadability scenarios. As a result, the feasible operating range is from the initial operating point to the SNB. Therefore, the critical parameters affecting the system stability limit are those that have the most significant impact on the feasible operating range (i.e., the location of the SNB). Figure 5.3 shows the participation factor analysis of the critical mode that causes the SNB. The participation factor values are distributed according to state variables belonging to different power components in the system, including SGs, VSCs, and network elements.

The state variables associated with SGs participate most in this mode, with a participation value of 0.98. Meanwhile, the state variables belonging to the VSC and the network have negligible participation in the critical eigenvalue. The participation factor analysis clearly demonstrates that the instability due to the SNB is caused by interaction among the power outputs of the critical SGs (i.e., G31 and G32 in this case). Therefore, the network impedances around G31 and G32, which have a significant impact on power flow around those critical SGs, strongly correlate

Table 5.1 The top five critical parameters affecting the feasible range with their correlation values for different control schemes

PV_{ac}	Parameter	Z_{6-31}	Z_{10-32}	Z_{3-4}	Z_{4-5}	Z_{9-39}
	Correlation value	– 1.00	– 0.95	– 0.80	– 0.75	– 0.70
PQ	Parameter	Z_{1-2}	Z_{6-31}	ω_{plli}	k_{pQi}	Z_{2-3}
	Correlation value	– 0.66	– 0.60	– 0.62	0.15	– 0.14

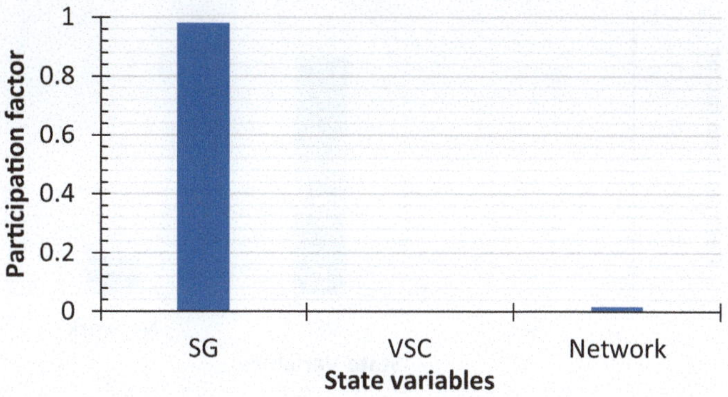

Fig. 5.3 Participation factor analysis of SNB

with the SNB values. Consequently, line impedances Z_{6-31}, Z_{10-32}, and Z_{3-4} can be considered critical parameters affecting system loadability. In contrast, the values of the calculated correlation indices between the control parameters of the VSC and SNBs are close to zero, as they have negligible impact on the power outputs of the key generators.

Table 5.1 illustrates that the critical parameters identified using the VSC with PQ control are significantly different from those identified when using PV_{ac} control. Unlike PV_{ac} control, it is shown that, with PQ control, the VSC control parameters correlate strongly with the feasible operating range. The respective correlation values of ω_{plli} and k_{pQi} are − 0.62 and 0.15. Furthermore, the impedances of transmission lines adjacent to G30 and G37 (i.e., Z_{1-2} and Z_{2-3}) correlate strongly with the feasible operating range with correlation values of − 0.66 and − 0.14, respectively. Additionally, Z_{6-31} which has significant impact on the locations of SNBs is still considered the influential parameter affecting the feasible operating range in this case.

When the VSC operates with PQ control, HB_1 and HB_2 may both occur along the system loadability trajectories in some scenarios. Based on the participation factor analysis results, HB_2 is due to the interaction between the outputs of SGs, which shares a similar instability mechanism with SNB. However, the key SGs affecting the locations of HB_2 are G30 and G37. Therefore, the impedances around G30 and G37 are also revealed to be critical parameters that have a significant impact on the feasible operating range in this case. This is confirmed by the correlation results shown in Table 5.1, where Z_{1-2} and Z_{2-3} have high correlations with the feasible operating range. The participation factor analysis of HB_1 that occurs under PQ control is illustrated in Fig. 5.4.

It can be seen that the state variables associated with the VSC participate most in the critical mode, followed by the state variables representing the network. As shown in Fig. 5.4, the participation factor values of these state variables of the VSC and the network are 0.86 and 0.14, respectively. The state variables belonging to the SGs have negligible impact in the critical mode, with a value of 0.0012. This shows that

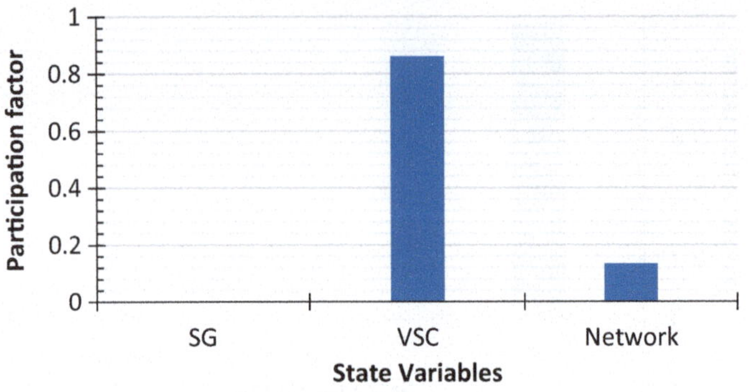

Fig. 5.4 Participation factor analysis of HB_1

HB_1 represents the interaction between the VSC and the network. Consequently, the correlation assessment under PQ control shows that VSC control parameters (i.e., ω_{plli} and k_{pQ}) have a significant impact on the HB_1 value and can further influence the feasible range along the loadability curve.

The study clearly shows the impact of VSC controller selection on the influential parameters identified. The VSC with PQ control is more likely to interact with the network and cause instability. Hence, these adverse interactions have significant impacts on the system's feasible range along the system loadability curve. This results in identifying different critical parameters using various VSC control schemes. Apart from that, this case shows that the system may become unstable via a small-disturbance instability (i.e., HB_1 due to interactions) before experiencing voltage instability (i.e., SNB or HB_2 due to the interaction between SGs' output), particularly when the PQ control is adopted. Hence, it is necessary to consider multiple types of stability simultaneously when assessing the system loadability.

5.2.2.2 The Weak-Grid Case

The work in [10] shows that the interaction between the VSC and the network will become more significant under weak-grid conditions and may further deteriorate system stability. Hence, in the weak-grid case, the initial value of the network impedance in the vicinity of the VSC is increased. Specifically, the impedance of lines 28–39, 26–28, and 26–29 is increased by a scaling factor of 1.5 for PV_{ac} control. For PQ control, the top five influential parameters identified in the weak-grid condition are the same as those from PQ control in the base case shown in Table 5.1. Hence, for brevity, the results of PQ control are not included in the weak-grid case.

For the weak-grid case, all bifurcations, including HB_1, HB_2, and SNB can occur along the system loadability trajectory. The instability mechanisms of different bifurcations are the same as those discussed in the base case (Sect. 5.2.2.1). Table 5.2 summarises the critical parameters affecting the feasible range considering PV_{ac}

Table 5.2 The top five critical parameters affecting the feasible operating range with their correlation value

PV_{ac}	Parameter	ω_{plli}	Z_{1-2}	Z_{6-31}	k_{pQi}	Z_{2-3}
	Correlation value	– 0.60	– 0.60	– 0.43	0.24	– 0.22

control. Similar to the PQ control results from Table 5.1, the impedance Z_{6-31} is recognised as a critical parameter. When the VSC is connected to a weak grid, the interaction between the VSC and the network, represented as HB_1, may deteriorate the system stability and affect the system loadability. Hence, the VSC control parameters ω_{plli} and k_{pQi} will have high correlation values with the feasible range, with values of – 0.6 and 0.24, respectively. Based on the participation factor analysis, the impedances around G30 and G37, i.e., Z_{1-2} and Z_{2-3}, correlate strongly with the location of HB_2. These parameters are also identified as influential factors affecting the system loadability in Table 5.2. In general, Table 5.2 identifies the same influential parameters as those shown in Table 5.1 when VSC is installed with PQ control

Combined with the results from Table 5.1, the sensitivity analysis performed in this section clearly shows that grid strength has a significant effect on the ranking of parameters based on their impact on system loadability. As is shown in Table 5.1, when a VSC with PV_{ac} is connected to a stiff network, the instability along the system trajectory is caused by interaction among the SG outputs. Therefore, the line impedances around the critical SGs can be considered influential parameters affecting system loadability and the control parameters of the VSC have a negligible impact on system loadability. If the VSC is connected to a weak grid, as is shown in Table 5.2, the interaction between the VSC and the network becomes significant and may further degrade the system loadability. Consequently, VSC control parameters (i.e., ω_{plli} and k_{pQi}) can affect system loadability and are, therefore, identified as influential parameters.

5.2.3 Comparison of Different Control Schemes

The methodology developed in the previous section was used to identify influential system parameters affecting system loadability. The proposed method in this section is used to assess the performance of different control schemes under various loadability scenarios, and hence to identify the VSC control that provides the largest loading margin. The same initial grid scenarios are used for both the PQ and PV_{ac} controls to ensure the results from the different VSC control schemes are comparable.

The typical loadability curve shown in Fig. 5.2 indicates that the feasible operating range for the system to operate is from HB_1 to HB_2. The value of the feasible range can be calculated by subtracting the location of HB_1 from the location of HB_2. Figure 5.5 shows how the value range of the feasible range changes as the scaling

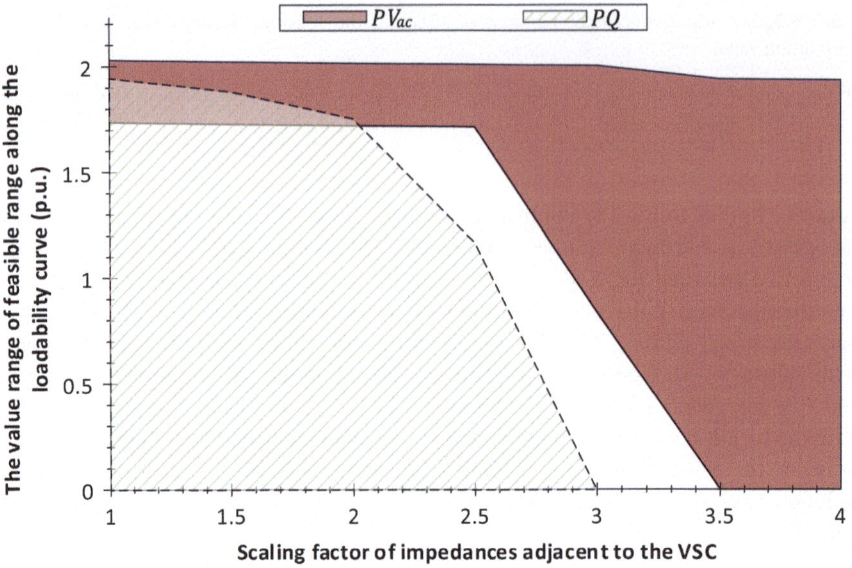

Fig. 5.5 Feasible operating ranges for different control schemes when one VSC is connected. (The feasible ranges are coloured in red for PV_{ac} and yellow for PQ. The feasible ranges are surrounded by a solid line for PV_{ac} and a dashed line for PQ)

factor of adjacent VSC impedance increases under different control schemes. The feasible ranges for PV_{ac} and PQ are shown in red (with a solid border) and green (with a dashed border), respectively.

When the VSC operates with PV_{ac} control, the maximum of the value range of the feasible range represents the loadability scenario where the entire loadability trajectory is stable (i.e., only SNB occurs along the curve). The upper limit of the value range gradually decreases from 2.03 to 1.94 p.u. with increases in the VSC adjacent impedances. The lower limit of the value range of the feasible range decreases at a similar rate when the scaling factor is below 2.5. When the impedance scaling factor increases above 2.5, interactions between the VSC and the network (shown as HB_1) begin to degrade the system and reduce the feasible range. Consequently, the lower limit of the value range of the feasible range decreases at a higher rate, from 1.71 p.u. and achieves 0.0 p.u. when the scaling factor reaches 3.5.

The maximum and minimum of the varying range of feasible ranges shown in Fig. 5.5 represent the optimal and worst loadability scenarios, respectively. The changes in the grid strength (the VSC adjacent impedance) affect the optimal and worst loadability curve simultaneously, as can be seen where the scaling factor is between 1 and 2.5. However, adverse interactions will affect the worst loadability scenario first, resulting in a corresponding reduction in the minimum value range of the feasible range. This is confirmed by observing the feasible range changes when the scaling factor increases from 2.5 to 3.5. If the impedance adjacent to the VSC further increases, the interactions begin to affect the optimal loadability scenario and thereby reduce the upper limit of the varying range of the feasible range.

When using the PQ control, the minimum of the value range of the feasible range is unaffected by the scaling factor and remains at 0.0 p.u. This occurs because, under the PQ control, the loadability scenarios include the unstable loadability curve sections for which there is no available feasible range. As discussed in Sect. 5.2.2.1, when the VSC operates with PQ control, adverse interactions occur even when the impedances adjacent to the VSC are at their nominal values. The maximum of the value range under PQ control decreases from 1.94 to 0.0 p.u at a faster rate than under PV_{ac} control.

Although the two controllers have overlapping feasible ranges when the impedance scaling factor is below 2.1, the PV_{ac} control can provide a larger feasible operating range than the PQ, and is less likely to cause system instability due to interactions. Additionally, the PV_{ac} control can operate over a wider range of loadability scenarios—particularly when connected to a weak grid and the VSC adjacent impedance is high. Hence, the PV_{ac} control scheme provides a wider feasible range and hence can be used to maintain the system stability when only one VSC is integrated into the system.

5.3 Application and Analysis of a System Containing Multiple VSCs

The developed method is used in this section to identify key system parameters and the VSC control scheme that provide the longest feasible range when multiple VSCs are connected within a system. Firstly, a similar process to that in which one VSC is connected (see the previous Sect. 5.2) is performed to identify critical parameters based on their impact on the system's feasible range. However, by considering a system with multiple VSCs integrated, it is possible to identify the critical parameters affecting instability due to the interaction between the VSCs via the AC network. The partial Spearman correlation between system parameters and the feasible operating range will be calculated for all parameters to select those showing a strong interdependency. Following this, the assessment of various VSC control schemes over a wide range of system operating conditions will be undertaken to identify the VSC controllers that offer the largest loading margin in power systems with multiple VSCs.

5.3.1 Modification to the Test Network

The modified IEEE 39-bus system used in this section is shown in Fig. 5.6, which shows two VSCs integrated on Buses 2 and 25 instead of the original SG 30 and SG

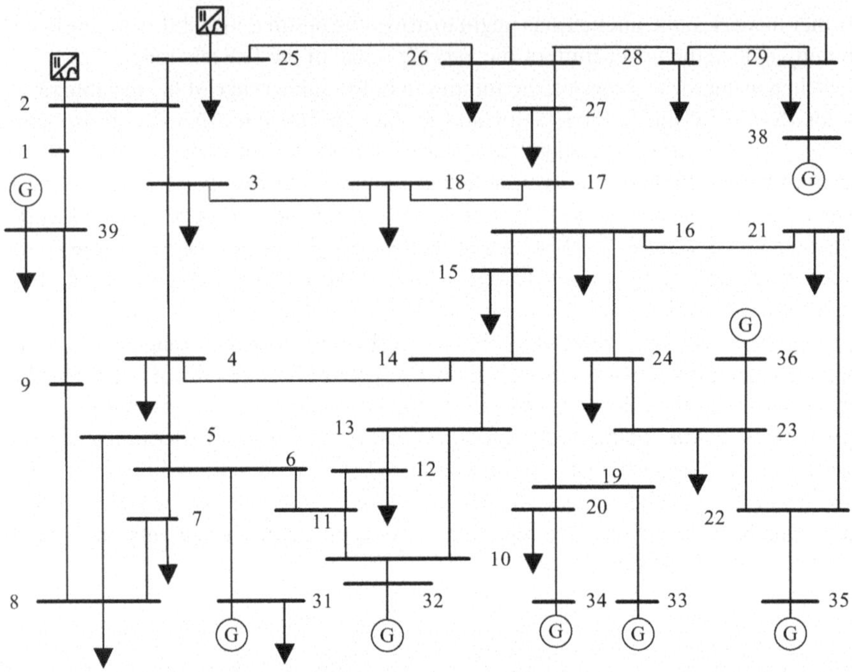

Fig. 5.6 The IEEE 39-bus system modified for two connected VSCs

37. To exacerbate the interaction, the active outputs are increased by a factor of 2.4 for Bus 2, and by 1.1 for the VSCs connected to Buses 25, respectively. Like the network modelling used in Sect. 5.2.1, this section dynamically models the entire network (i.e., the dynamic model developed in Chap. 2) to capture instability caused by interaction between the VSCs.

5.3.2 *Identification of Influential Parameters*

In this section, the combinations of control schemes of the two VSCs considered are the PV_{ac}–PV_{ac}, PQ–PV_{ac}, and PQ–PQ. It should be noted that the critical parameters identified in the base case of this section are the same as those identified in the previous base case (Sect. 5.2.2.1), in which a single VSC with the PV_{ac} control is integrated into the system. This is because, in the base case, where the grid is stiff, most loadability curves are entirely stable and only the SNB occurs along the trajectory. Based on the participation factor analysis shown in Fig. 5.3, SNBs are mainly due to the interactions between SG outputs. Under such circumstances, VSC control schemes have a negligible impact on the locations of SNBs. Therefore, the top five critical parameters are identical for each case and are the same as the results shown in Sect. 5.2.2.1 regardless of which controller combination is used. For the sake of brevity, the results of the base case are not included in this section. Therefore, this section focuses mainly on the weak-grid case, in which instability caused by interaction between the VSCs occurs along the system loadability trajectory.

To exacerbate the interaction between the VSCs in a similar way as in Sect. 5.2.2.2, the impedance of Lines 1–2, 2–25, 25–26 and 2–3 around the VSCs is increased in the weak-grid case to reduce the SCR seen by the VSCs. Specifically, the impedance of the transmission lines is increased by a factor of 1.4 for the PQ–PQ control, 2.4 for the PQ–PV_{ac} control, and 4.6 for the PV_{ac}–PV_{ac} control.

5.3.2.1 The Weak-Grid Case

In this case, HB_1, HB_2, and SNB can all exist on the system loadability curves for all VSC control schemes. According to the participation factor analysis, HBs are caused by interactions between the VSCs, while the SNB is due to the interaction between SG outputs. Table 5.3 summarises the priority ranking of the system parameters based on their influence on the feasible range for different VSC controls. As illustrated in Table 5.3, all three VSC controllers indicate that the impedance adjacent to the critical SG (e.g., Z_{6-31}) correlates strongly with the feasible operating range, since it has a significant impact on the location of SNBs. Furthermore, the impedance of transmission lines in the vicinity of the VSCs (e.g., Z_{2-3}) and VSC control parameters (e.g., ω_{plli}), which can influence the locations of HBs, are also strongly correlated with the feasible operating range. Taking the results of the PQ–PQ combination as an example, the value of the calculated correlation indices of Z_{6-31}, Z_{2-3}, and ω_{plli} are − 0.82, − 0.31, and − 0.26, respectively. From these results, it can be concluded that the interaction between the VSCs via the AC network is strongly affected by the impedance around the VSCs and the VSC control parameter (i.e., ω_{plli}), and that the interaction between SGs is affected by the adjacent impedance. Therefore, these parameters are considered critical parameters affecting the feasible operating range of the system.

The sensitivity study conducted on the multiple VSCs case again shows the impact of the grid strength on the critical parameters identified. When the VSCs are connected to a stiff network, the system has an entirely stable loadability curve and

Table 5.3 The top five critical parameters affecting the feasible operating range with their correlation value considering different VSC control schemes

PQ–PQ	Parameter	Z_{6-31}	Z_{1-2}	Z_{2-3}	Z_{10-32}	ω_{plli}
	Correlation value	− 0.82	− 0.45	− 0.31	− 0.30	− 0.26
PQ–PV_{ac}	Parameter	Z_{6-31}	Z_{1-2}	Z_{10-32}	Z_{4-5}	ω_{plli}
	Correlation value	− 0.89	− 0.36	− 0.35	− 0.18	− 0.18
PV_{ac}–PV_{ac}	Parameter	Z_{6-31}	Z_{1-2}	Z_{10-32}	ω_{plli}	Z_{8-9}
	Correlation value	− 0.84	− 0.40	− 0.30	− 0.22	− 0.19

the influential parameters are the impedances around key generators. However, under weak-grid conditions, the interaction between VSCs will affect the system loadability and reduce the feasible range. For such scenarios, the VSC control parameters and VSC adjacent impedances are recognised as key parameters. Additionally, this case also demonstrates the importance of considering multiple types of stability simultaneously during a loadability assessment as the system may encounter small-disturbance instability before meeting voltage instability.

5.3.3 Comparison of Different Control Schemes

This section compares the performance of different VSC control schemes over a wide range of loadability scenarios using the modified network shown in Fig. 5.7. The study will identify the VSC controllers that provide the largest loading margin in power systems where multiple VSCs are closely connected. The same initial grid scenario is used in all three control schemes to ensure the results are comparable.

Figure 5.7 displays the feasible range of different controllers as the scaling factor of VSC adjacent impedances increases. The feasible ranges for the PV_{ac}–PV_{ac}, PQ–PV_{ac}, and PQ–PQ are shown in red (with a solid border), yellow (with a dashed border) and blue (with a dotted border), respectively.

Figure 5.7 illustrates that the PV_{ac}–PV_{ac} control has a feasible range similar to that shown in Fig. 5.5 when PV_{ac} is used. The upper limit represents the scenario where only SNBs occur along the system loadability curves. The maximum feasible range decreases slowly and steadily from 2.01 to 1.76 p.u. at a relatively constant speed as the scaling factor increases from 1 to 6. The lower limit shows a stable loadability curve for the impedance scaling factor below 3.4. Specifically, only SNBs occur along the system loadability trajectory when the scaling factor is lower than 3.4. For the scaling factor above 3.4, adverse interactions between VSCs start to affect the system loadability and reduce the minimum feasible range. When the scaling factor is smaller than 3.4, the minimum of the value range gradually decreases from 1.73 to 1.60 p.u. at a similar rate to the maximum of the value range; when the scaling factor is larger than 3.4, it decreases from 1.60 to 0.0 p.u. much more rapidly.

For the PQ–PV_{ac} controller combination, the upper limit of the value range of the feasible range represents a loadability trajectory where HBs disappear and only SNBs occur for the scaling factor below 3.2. Further increases in the impedances adjacent to VSCs cause adverse interactions to begin affecting the maximum value range of the feasible range. Hence, the upper limit of the value range decreases rapidly from 1.80 to 0.0 p.u. when the scaling factor reaches 4.8. The minimum feasible range of the PQ–PV_{ac} control represents the loadability curve along which only an SNB occurs if the impedance scaling factor is below 1.8. Compared to the PV_{ac}–PV_{ac} combination, the PQ–PV_{ac} control is more likely to couple through the network and shrink the system's feasible region. As a result, the coupling effect between the VSCs (represented as HB) drives the lower limit of the value range rapidly to 0.0 as the scaling factor increases above 1.8.

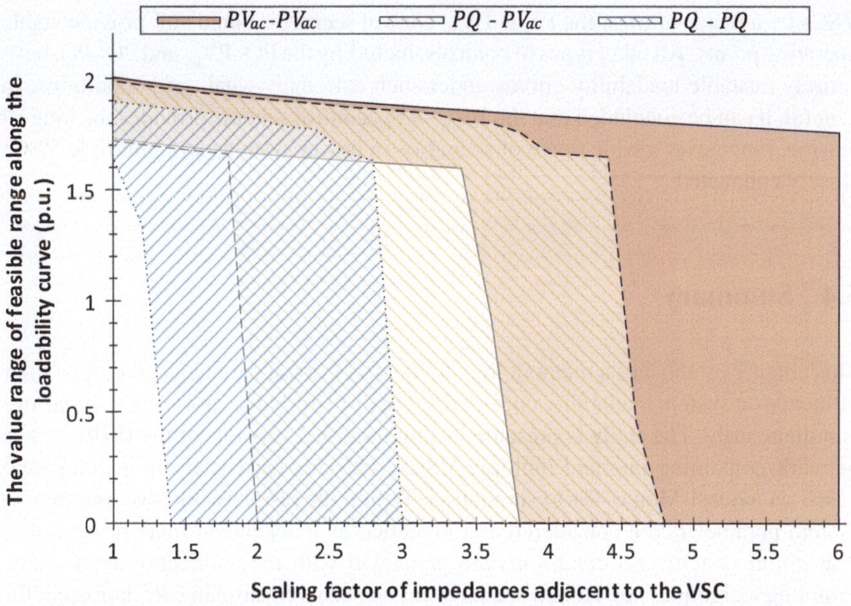

Fig. 5.7 Feasible operating ranges for different control schemes when two VSCs are connected. (The feasible ranges are coloured in red for PV_{ac}–PV_{ac}, yellow for PQ–PV_{ac} and blue for PQ–PQ. The feasible ranges are surrounded by a solid line for PV_{ac}–PV_{ac}, a dashed line for PQ–PV_{ac} and a dotted line for PQ–PQ)

The maximum of the value range for the PQ–PQ combination shows a similar loadability scenario to the other two types of controllers in which only an SNB occurs at the tip of the curve, when the scaling factor is below 2.5. Further impedance increases cause interactions between the VSCs, resulting in the reduction of the maximum of the value range of the feasible range for the PQ–PQ control scheme. As the impedance scaling factor increases from 2.5 to 3.0, the maximum of the value range shrinks from 1.77 to 0.0 p.u. For the lower limit of the varying range, the adverse coupling effect between VSCs quickly reduces the feasible range from 1.65 to 0.0 p.u. as the scaling factor increases from 1.0 to 1.4.

Figure 5.7 shows how interactions start to affect the minimum value range for the worst loadability scenarios when the scaling factor is above certain values that are different from the three control schemes. Specifically, the minimum of the value range is affected when the scaling factor increases above 3.4 for PV_{ac}–PV_{ac}, 1.8 for PQ–PV_{ac}, and 0.0 for PQ–PQ. It can be established that VSCs with PV_{ac}–PV_{ac} control have a lower potential to interact with each other and cause system instability. However, Fig. 5.7 also indicates that when the scaling factor is below 1.8, the feasible range of the PQ–PV_{ac} and PV_{ac}–PV_{ac} controls almost overlap. This shows that the PQ–PV_{ac} control has similar performance to the PV_{ac}–PV_{ac} control when VSCs are connected to a relatively strong AC grid. The results shown in Fig. 5.7 reveal that even under higher impedance conditions (e.g., increasing the impedance around the

VSCs by a factor of 6), some PV_{ac}–PV_{ac} control scenarios could still provide stable operating points. All other types of controls, including the PQ–PV_{ac} and PQ–PQ, have entirely unstable loadability curves under such extremely weak-grid conditions. In general, it can be concluded that the PV_{ac}–PV_{ac} control scheme provides the longest feasible range over a wide range of scenarios in power systems with multiple VSCs closely connected.

5.4 Summary

This chapter establishes a methodology to identify critical parameters based on their influence on system loadability considering small-disturbance and voltage stabilities simultaneously. The study is presented using modified versions of the IEEE 39-bus network containing one and multiple VSCs. The proposed analysis is completed based on a quasi-Monte–Carlo simulation. Within the test network, uncertainties in system parameters are considered and modelled as independent random variables. The initial system grid conditions are generated with the application of a quasi-sampling technique (i.e., Halton sequence). The partial Spearman correlation coefficient is employed to quantify the impact of each parameter on the critical points along the system loadability curves to determine influential system parameters affecting the system loadability. Alternatively, the developed methodology can also be used to evaluate the performance of different VSC controllers over wide-ranging loadability scenarios to establish which controllers are able to maintain stable system operation.

The comprehensive study conducted across a wide range of scenarios has clearly illustrated the influence of different grid strengths and the selection of VSC controllers on the influential parameters identified. The results show that under nominal grid conditions, the instability along the system loadability trajectory is commonly caused by the power output of the SGs. Therefore, the impedance of the transmission lines adjacent to these critical SGs is considered as an influential parameter affecting the system loadability. The control parameters of SGs and VSCs under normal circumstances have a negligible impact on the system loadability. However, under weak-grid conditions, the network impedances in the vicinity of the VSCs and the control parameters of the VSCs are strongly correlated with the critical bifurcation points representing the interaction between the VSCs and the network. These parameters will thus impact the system's feasible operating range further and are identified as critical parameters under weak-grid conditions. The identification of critical parameters that influence system loadability can help the system operator to optimise the system operation with minimal computational resources. Additionally, these studies have clearly shown that the system may become unstable through small-disturbance instability (i.e., adverse interactions represented as HBs) before encountering the voltage operational limits (represented as SNBs). This demonstrates the necessity of considering both small-disturbance and voltage stability when assessing the system loadability.

In this chapter, the proposed method is also utilised to conduct a comprehensive evaluation of the performance of various VSC controllers across diverse loadability conditions. This clearly demonstrates that when one VSC is connected to the system, the PV_{ac} control is likely to provide a wider feasible operating range. Therefore, the PV_{ac} control is recommended for systems with a single VSC installed. For power systems with multiple VSCs closely connected, the PV_{ac}–PV_{ac} control provides the widest feasible range. It should be noted that the PV_{ac}–PV_{ac} control has the potential to provide stable operating points under extremely weak-grid conditions. Using the recommended VSC control will help the system operator to maintain system stability under various loadability conditions.

The method developed in this chapter assesses the system loadability with the simultaneous consideration of small-disturbance and voltage stability based on bifurcation theory. However, to capture different types of instabilities simultaneously, the proposed approach requires a dynamic representation of the entire network. This means that the entire network should be modelled with ODEs in the developed method. For studies performed on a small-size test system, this might be acceptable, given that the number of state variables included is not particularly large. However, when a large power system is considered, the inclusion of all network dynamics indicates a significant increase in the dimension of the system state-space matrix and will ultimately lead to a prohibitively long simulation time. The next chapter will present an analysis framework based on hybrid network modelling, which can be used to accurately capture system dynamic characteristics using a greatly reduced size of the system matrix.

References

1. Qi B, Hasan KN, Milanovic JV (2019) Identification of critical parameters affecting voltage and angular stability considering load-renewable generation correlations. IEEE Trans Power Syst 34(4):2859–2869
2. Fatah Mochamad R, Preece R, Hasan KN (2022) Probabilistic multi-stability operational boundaries in power systems with high penetration of power electronics. Int J Electr Power Energy Syst 135:107382
3. Hatziargyriou N et al (2020) Stability definitions and characterization of dynamic behavior in systems with high penetration of power electronic interfaced technologies. IEEE Power Energy Soc USA Tech Rep TR-0200 (4230–46)-3
4. Zhou Y, Ajjarapu V (2005) A fast algorithm for identification and tracing of voltage and oscillatory stability margin boundaries. Proc IEEE 93(5):934–946
5. Gomes S, Martins N, Parreiras TJMA (2017) Computing the closest small-signal security boundary in the control parameter space for large scale power systems. Electr Power Syst Res 149:10–18
6. Veloza OP, Santamaria F (2016) Analysis of major blackouts from 2003 to 2015: classification of incidents and review of main causes. Electr J 29(7):42–49
7. Arani MFM, Mohamed YARI (2017) Analysis and performance enhancement of vector-controlled VSC in HVDC links connected to very weak grids. IEEE Trans Power Syst 32(1):684–693
8. Zhang L, Nee HP, Harnefors L (2011) Analysis of stability limitations of a VSC-HVDC link using power-synchronization control. IEEE Trans Power Syst 26(3):1326–1337

9. Mochamad RF, Preece R (2020) Assessing the impact of VSC-HVDC on the interdependence of power system dynamic performance in uncertain mixed AC/DC systems. IEEE Trans Power Syst 35(1):63–74
10. Zhou JZ, Ding H, Fan S, Zhang Y, Gole AM (2014) Impact of short-circuit ratio and phase-locked-loop parameters on the small-signal behavior of a VSC-HVDC converter. IEEE Trans Power Deliv 29(5):2287–2296

Chapter 6
Analysing System Loadability with Multiple VSCs Using a Hybrid Model

The previous chapter has demonstrated the need to consider small-disturbance and voltage stability simultaneously when assessing a future power system with multiple VSCs. However, the network in Chap. 5 is represented by a dynamic network model to capture those newly emergent power system issues related to power electronic-based devices. Specifically, the entire network, including all transmission lines, buses, and loads, is modelled dynamically with ODEs to capture the higher-frequency oscillations. Dynamic modelling of an entire system with ODEs would result in a prohibitively large number of state variables to be included and, consequently, a high-dimensional system Jacobian matrix. This, in turn, would ultimately lead to an excessively long simulation time. It is impractical to model all the network dynamics of real large power systems due to the modelling complexity [1]. Additionally, the necessity of including a dynamic representation of all power elements within the model remains unclear [2].

The use of hybrid modelling proposed in [1] provides a potential solution for capturing the higher-frequency oscillations caused by the VSCs when analysing the small-disturbance stability of large power systems accurately without incurring the computational complexity and burden of the full-dynamic network representation. In the hybrid model, the network components in the vicinity of power converters are modelled with ODEs, while the rest of the network modelled using algebraic equations. Therefore, the hybrid model significantly reduces the dimension of the system Jacobian matrix required to complete the system small-disturbance stability analysis, whilst still accurately providing the system stability information. Similar modelling ideas have been applied in other areas. The work described in [3], for example, divides a system into dynamic and static parts to analyse the dynamics of a VSC–HVDC link.

Y. Chen, *Stability Assessment of Power Systems with Multiple Voltage Source Converters*, Springer Theses, https://doi.org/10.1007/978-3-031-63095-8_6

Within the existing research, hybrid modelling has typically been adopted for the purpose of preserving converter-related critical dynamics. Although these newly emergent power electronic-related issues are important for future power system stability, the dynamics of other non-converter equipment which may also cause system instabilities are neglected in the hybrid models developed in the previous literature.

The region of the network modelled with ODEs, known as the dynamic area, has a significant impact on stability assessments, particularly with regard to the accuracy of the results. Hence, the selection of the dynamic area of the network is a vital step during the construction of the network hybrid model. Although there has been some discussion about the impact of dynamically modelling different portions of the network in [2], a methodology has not yet been established that provides a clear, scientific definition to determine the network portion that should be modelled dynamically in the hybrid model.

This chapter represents the sixth contribution of the thesis: an analysis framework based on the hybrid model for the system loadability assessment with simultaneous consideration of small-disturbance and voltage stability. The dynamic network region modelled with ODEs in the proposed hybrid model is defined based on eigenvalue sensitivity and participation factor analyses. Following a demonstration on a small test system, the developed framework is then extensively tested on a large IEEE 39-bus network. The different system instabilities caused by network dynamics alone, the interaction between the VSC and the network, and the interactions between multiple VSCs via AC networks are identified and assessed using the analysis framework. Each scenario compares various types of system information from different models, including system stability assessment, the number of state variables and the simulation times. The comparative studies clearly demonstrate that the proposed hybrid model is capable of producing accurate system stability results with a lower computational burden.

6.1 Hybrid Network Modelling

Within the hybrid model, the entire AC network is separated into two regions: the dynamic and static parts. The dynamic network area is modelled with ODEs considering the network LC dynamics, and the static network region is represented by sets of algebraic equations. The ODE-represented area (dynamic area) is identified by the methodology developed in Sects. 6.2.1–6.2.2.

6.1.1 Static Area of the Network

The completed integrated hybrid model is shown in Fig. 6.1. The upper portion of Fig. 6.1 depicts the integration of the network's static area. In general, the static

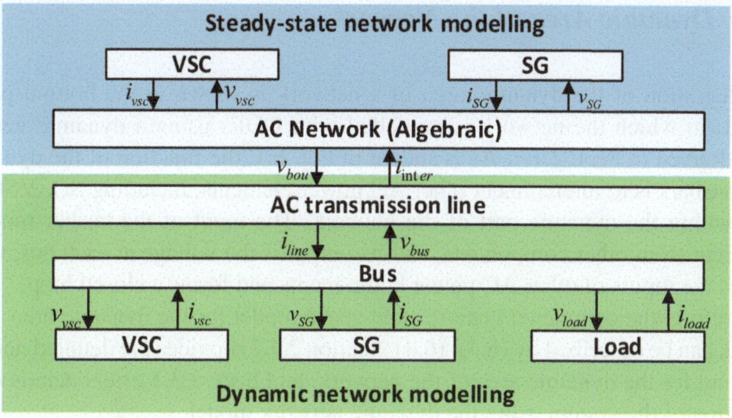

Fig. 6.1 The complete hybrid model. © 2024 IEEE, reprinted, with permission, from Chen et al. [4]

part of the network is modelled with algebraic equations using the static network modelling developed in Sect. 2.1.6. The output currents of the SGs and VSCs match the injected currents in the static network model. The network model provides the voltages at each busbar, which serve as inputs to the VSCs and SGs.

Within the model of the static area of the network, the current injections from the VSCs and SGs are fed into the network, and the network then feeds back the voltages. The generic state-space model of the static part of the grid can be represented by (6.1)–(6.2). The detailed network modelling of the static area of the network is discussed in Sect. 2.1.6 and details of the system integration using the static network modelling are included in Sect. 2.3.1.

$$\frac{dx^{sta}}{dt} = A^{sta}x^{sta} + B^{sta}\left[i_{\mathrm{interDi,Qi}}\right] \tag{6.1}$$

$$v_{\mathrm{bouDi,Qi}} = C^{sta}x^{sta} \tag{6.2}$$

In (6.1)–(6.2), $i_{\mathrm{interDi,Qi}}$ represents a vector of the currents through the interconnected lines that connect the static and the dynamic areas of the network, in the *DQ* frame; $v_{\mathrm{bouDi,Qi}}$ is a vector of the voltages at the boundary of the static and the dynamic network portions, in the *DQ* frame; x^{sta} is the vector of the state variables of the power components included in the static area of the network (for example, the state variables of the SGs and VSCs in the static network area).

6.1.2 Dynamic Area of the Network

The integration of the dynamic area of a network is shown in the bottom part of Fig. 6.1, in which the network is modelled with ODEs using a dynamic network model derived in Sect. 2.1.7. As is shown in Fig. 6.1, the function of the dynamics of the busbars is to interconnect other AC power elements, including SG, VSC and loads, within the dynamic part of the network. The input of the busbar model is the current from other components, and its output is the voltage at each bus, which matches the inputs of other AC power components and forms a closed loop.

Therefore, the completed generic state-space model for the dynamic area of the network can be described by (6.3)–(6.4). Section 2.1.7 provides the detailed network modelling for the dynamic area of the network, and Sect. 2.3.1 offers details of the integration of the system using the dynamic network model.

$$\frac{dx^{dyn}}{dt} = A^{dyn}x^{dyn} + B^{dyn}\left[u_{bouDi,Qi}\right] \tag{6.3}$$

$$i_{interDi,\,Qi} = C^{dyn}x^{dyn} \tag{6.4}$$

In (6.3)–(6.4), x^{dyn} is the vector of the state variables of the power components included in the dynamic area. This includes the state variable from the SGs, VSCs, transmission lines, and loads in the dynamic area of the network.

6.1.3 Integration

The overall hybrid model shown in Fig. 6.1 is completed by connecting the static area modelled by (6.1)–(6.2) and the dynamic area modelled by (6.3)–(6.4). Specifically, the output of the dynamic network modelling, i.e., currents across the interconnected lines that connect the static and the dynamic regions of the network, matches the input of the static network modelling. The output of (6.1)–(6.2) provides the voltages at boundary buses, which present the input of (6.3)–(6.4). In general, the currents flowing through the transmission lines connecting the static and dynamic areas of the network, and the voltages at the boundary between these areas, act as a junction that integrates the two different modelling parts of the network. Hence, the dynamic and static parts of the network are integrated together to form a completed closed loop.

6.2 Proposed Methodology

This section will introduce the methodology to identify the dynamic area of the network within the hybrid model which should be modelled with ODEs. Various techniques, including eigenvalue sensitivity and participation factor analysis, have been incorporated within the proposed methodology. The analytical framework based on the hybrid modelling scheme can be used to assess loadability limits in systems accurately with multiple VSCs considering different instability mechanisms, whilst significantly reducing the computational burden. The continuation method is used to fully track the system loadability trajectory with consideration of multiple bifurcations (i.e., multiple instability types).

The developed approach can be divided into four stages, which are detailed below:

Step 1: A full-dynamic model is created in which the entire network is modelled with ODEs. The critical mode is determined from the full-dynamic system matrix based on the eigenvalue sensitivity.

Step 2: The key state variables are identified based on participation factor analysis of the critical mode. The key power components (related to key state variables) and their adjacent network components are selected to be modelled dynamically.

Step 3: The hybrid model, which includes dynamic and static areas, is created based on the hybrid modelling approach proposed in Sect. 6.1.

Step 4: The system loadability curve is calculated using the continuation method and the developed hybrid model. The system loadability assessment is performed using bifurcation analysis.

6.2.1 Determination of Critical Modes

The first stage in the proposed method is to determine the critical eigenvalue that first crosses the imaginary axis, based on eigenvalue sensitivity. In general, the trajectory of an eigenvalue can be considered a function of uniform acceleration movement, with the first- and second-order eigenvalue sensitivities mapped to velocity and acceleration concepts from kinematics [5, 6]. The eigenvalue with the shortest time to instability (i.e., the eigenvalue expected to reach the imaginary axis first) will be considered the critical mode.

As discussed in Chap. 2, the dynamic system state matrix A_{sys} is obtained from (2.56) when the power system is modelled by a set of DAEs. The obtained A_{sys} is a reduced matrix, and hence, it is hard to directly derive eigenvalue sensitivity based on it. In this study, eigenvalue sensitivity will be directly calculated from an augmented (unreduced) system matrix.

6.2.1.1 First-Order Eigenvalue Sensitivity with Respect to a Bifurcation Parameter

To determine the eigenvalue sensitivity, a power system represented by a set of DAEs can be rearranged into its state-space form (2.51). For a given eigenvalue λ_i, the corresponding left and right eigenvectors are w_i and v_i, therefore, (6.5)–(6.6) can be stated.

$$\begin{bmatrix} \lambda_i v_i \\ 0 \end{bmatrix} = \begin{bmatrix} f_x & f_y \\ g_x & g_x \end{bmatrix} \begin{bmatrix} v_i \\ v_{yi} \end{bmatrix} \tag{6.5}$$

$$\begin{bmatrix} \lambda_i w_i^T & 0 \end{bmatrix} = \begin{bmatrix} w_i^T & w_{yi}^T \end{bmatrix} \begin{bmatrix} f_x & f_y \\ g_x & g_x \end{bmatrix} \tag{6.6}$$

In (6.5)–(6.6), v_{yi} and w_{yi}^T are the expanded parts of the corresponding eigenvector, since an augmented matrix is used.

If the system augmented matrix is $\tilde{A} = \begin{bmatrix} f_x & f_y \\ g_x & g_x \end{bmatrix}$, the corresponding left and right eigenvectors of the augmented matrix are $\tilde{w}_i = \begin{bmatrix} w_i \\ w_{yi} \end{bmatrix}$ and $\tilde{v}_i = \begin{bmatrix} v_i \\ v_{yi} \end{bmatrix}$. Equations (6.5) and (6.6) can then be simplified to (6.7) [7].

$$\begin{cases} \tilde{A}\tilde{v}_i = \lambda_i \tilde{I} \tilde{v}_i \\ \tilde{A}^T \tilde{w}_i = \lambda_i \tilde{I} \tilde{w}_i \end{cases} \tag{6.7}$$

In (6.7), $\tilde{I}$ is an identity matrix and all other positions have values of 0.

The first-order eigenvalue sensitivity with respect to the bifurcation parameter μ is then determined, as in (6.8).

$$\frac{\partial \lambda_i}{\partial \mu} = \tilde{w}_i^T \frac{\partial \tilde{A}}{\partial \mu} \tilde{v}_i \tag{6.8}$$

6.2.1.2 Second-Order Eigenvalue Sensitivity with Respect to a Bifurcation Parameter

It is also possible to extend the sensitivity analysis to determine second-order effects. Differentiating (6.7) with respect to the bifurcation parameter μ gives (6.9) [6].

$$\begin{cases} \left(\tilde{A} - \lambda_i \tilde{I}\right) \frac{\partial \tilde{v}_i}{\partial \mu} = \tilde{M} \tilde{v}_i \\ \left(\tilde{A}^T - \lambda_i \tilde{I}\right) \frac{\partial \tilde{w}_i}{\partial \mu} = -\tilde{N} \tilde{w}_i \end{cases} \tag{6.9}$$

In (6.9), $\tilde{M} = \frac{\partial \tilde{A}}{\partial \mu} - \frac{\partial \lambda_i}{\partial \mu}\tilde{I}$ and $\tilde{N} = \frac{\partial \tilde{A}^T}{\partial \mu} - \frac{\partial \lambda_i}{\partial \mu}\tilde{I}$. However, since $(\tilde{A} - \lambda_i \tilde{I})$ in (6.9) is singular, the determination of eigenvector sensitivity is not uniquely guaranteed. Additional conditions, where the right eigenvector sensitivity vector is orthogonal to the left eigenvector, and the left eigenvector sensitivity vector is orthogonal to the right eigenvector, are considered. Hence, two additional equations are added, as shown in (6.10).

$$\begin{cases} \tilde{w}_i^T \tilde{I} \frac{\partial \tilde{v}_i}{\partial \mu} = 0 \\ \tilde{v}_i^T \tilde{I} \frac{\partial \tilde{w}_i}{\partial \mu} = 0 \end{cases} \tag{6.10}$$

Unique solutions to the problem of left/right eigenvector sensitivity can be found by solving both (6.9) and (6.10). Differentiating (6.8) with respect to the bifurcation parameter μ gives the second-order eigenvalue sensitivity shown in (6.11) [6].

$$\frac{\partial^2 \lambda_i}{\partial \mu} = \frac{\partial \tilde{w}_i^T}{\partial \mu}\frac{\partial \tilde{A}}{\partial \mu}\tilde{v}_i + \tilde{w}_i^T \frac{\partial^2 \tilde{A}}{\partial \mu}\tilde{v}_i + \tilde{w}_i^T \frac{\partial \tilde{A}}{\partial \mu}\frac{\partial \tilde{v}_i}{\partial \mu} \tag{6.11}$$

6.2.1.3 Determination of Key Eigenvalue

The critical condition to maintain system stability within this research is to keep the real part of all eigenvalues negative, i.e., $\sigma < 0$, which is also adopted in [5, 8]. Consequently, the eigenvalue that first passes through the imaginary axis is considered the critical mode. Therefore, the real part of eigenvalue sensitivity indices developed in the previous section, i.e., $\frac{\partial \sigma_i}{\partial \mu}$ and $\frac{\partial^2 \sigma_i}{\partial \mu}$, will be adopted in this section to identify the critical mode.

From a kinematics point of view, for a given uniform acceleration a and an initial speed V_0, the time t required to finish a given path S can be expressed as (6.12).

$$S = V_0 t + \frac{1}{2}\mathrm{at}^2 \tag{6.12}$$

Similarly, if a system is initially stable, for any eigenvalue moving to the right-hand side, it can be assumed that the movement has uniform acceleration. This assumption has been adopted in previous research [5, 8] when identifying the most critical mode that first passes through the imaginary axis. Although under a very small number of cases, the movements of some eigenvalues are less consistent, the assumption of motion with constant acceleration represents the typical movements of the majority of eigenvalues. Based on more than 10^4 numerical simulations completed in this research, the above assumption is correct for over 90% of cases. The kinematics concepts in (6.12) can then be represented by eigenvalue sensitivity, with parameters mapping as shown in (6.13).

$$\sigma_i \to S, \quad \frac{\partial \sigma_i}{\partial \mu} \to V_0, \quad \frac{\partial^2 \sigma_i}{\partial \mu} \to a, \quad \mu \to t \tag{6.13}$$

Hence, (6.12) can be expressed as (6.14) with respect to eigenvalue sensitivity.

$$|\sigma_i| = \frac{\partial \sigma_i}{\partial \mu} \mu_{c\_i} + \frac{1}{2} \frac{\partial^2 \sigma_i}{\partial \mu} \mu_{c\_i}^2 \tag{6.14}$$

Therefore, the expected time (i.e., the change in the bifurcation parameter μ_{c_i}) required to reach the imaginary axis is determined as in (6.15) [9].

$$\mu_{c\_i} = \frac{-\frac{\partial \sigma_i}{\partial \mu} + \sqrt{\left(\frac{\partial \sigma_i}{\partial \mu}\right)^2 + 2\frac{\partial^2 \sigma_i}{\partial \mu}|\sigma_i|}}{\frac{\partial^2 \sigma_i}{\partial \mu}} \tag{6.15}$$

If the system is initially stable, the eigenvalue which has the smallest value of μ_{c_i} is considered the critical mode.

For initially unstable systems, $\sigma_i > 0$, and the eigenvalue would now be expected to move towards the left-hand side of the complex plane with μ increasing. Hence, the eigenvalue that has the largest value of μ_{c_i} will be considered as critical mode.

6.2.2 *Identification of Key Power Components*

After the critical eigenvalue is identified, the second stage of the proposed methodology is to identify the key power components that require dynamic modelling. The participation factor analysis technique discussed in Sect. 2.4.1 is used to determine this. A large participation factor value reveals a significant level of state variable participation in the corresponding mode [10].

With the critical mode determined as described in the previous section, the participation of each state variable with respect to the critical mode can be calculated. The state variables can then be ranked in descending order of their absolute participation factor values. The top state variables with high participation factors are the key state variables. Determining a threshold of what constitutes high participation is a subject that requires knowledge of the system being considered. In this section, all state variables with participation values greater than α times the highest value are considered key state variables, where α is a scalar factor, which, in this chapter, is 0.1.

The power components represented by the identified key state variables are considered as key components. Dynamic network representation is used for all portions of the network that include the key components, as well as for the transmission lines and buses immediately adjacent to the key components. For example, if a bus is identified as a key component, then that bus, as well as all the lines and devices connected to that bus, will be represented with dynamic models. Therefore, this step determines the dynamic and static areas of the network. The dynamic region of the network consists of all key components as well as network elements adjacent to the key components. The remaining parts of the network are recognised as the static area of the network, which is represented by algebraic equations.

6.2.3 Construct the Hybrid Model

Once the dynamic network area is identified, the rest of the network will be modelled with algebraic equations, and the hybrid model can be created based on the modelling method described in Sect. 6.1.

6.2.4 Complete the Loadability Assessment

The system loadability curve will then be calculated by applying the continuation method to the developed hybrid model [11]. The continuation method is selected in this research, as it can provide the most detailed information about the system trajectory with respect to multiple instability mechanisms. Other time-efficient techniques, such as optimisation-based analysis frameworks, developed in Chap. 3, could also be adopted if desired. Once the system loadability curve is obtained, the loadability assessment can be performed using bifurcation analysis.

It should be noted that the proposed methodology builds the hybrid model from the information provided by the full-dynamic model. This is a typical step in various approaches used to simplify a detailed system model. Take the dynamic reduction programme DYNRED, developed under the Electric Power Research Institute's (EPRI) project RP2447, as an example [12]. The network reduction programme consists of the following three steps to form a reduced system model: (1) Obtain information from the detailed system model; (2) Identify the power components and network area of interest; (3) Model the network area of interest with a detailed model, and represent the remainder of the system by a simplified model. The analysis framework developed in this chapter follows a similar network simplification process to build the hybrid model. Although the proposed approach requires a fully dynamic model at the initial operating point in which the entire system is modelled with ODEs, the proposed framework still reduces the computational burden, as it computes the entire system equilibrium trajectory when performing the system loadability assessment. The system loadability curve within this study typically contains 10^3–10^4 operating points. Therefore, compared to the fully dynamic model, the developed hybrid model with its lower-dimensional matrix is a computationally efficient tool for analysing the system loadability trajectory. This is further illustrated and supported by the numerical results presented later in the chapter.

Apart from that, it is worth noting that the developed analysis framework may not be applicable for scenarios where the key state variables identified in Sect. 6.2.2

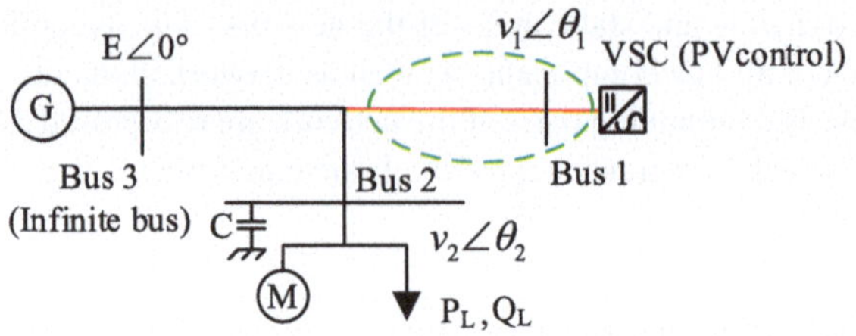

Fig. 6.2 The modified three-bus test system (The key elements are coloured in red, and the dynamic area is surrounded by a dashed curve). © 2024 IEEE, reprinted, with permission, from Chen et al. [4]

change significantly as the system loading increases. However, within the simulations carried out in this section, the key state variables that predominantly control the critical eigenvalue remain unchanged as the bifurcation increases.

6.3 Case-1: Three-Bus System

The previous sections developed an analysis framework to efficiently assess the system loadability limit considering different types of stabilities. The three-bus system in this section is used to illustrate the effectiveness of the proposed modelling approach.

6.3.1 *Modification to the Test Network*

The modified three-bus system is illustrated in Fig. 6.2.

Compared with the original three-bus system, the SG at Bus 1 is replaced with a VSC operating in PV_{ac} mode [13]. The VSC parameters are given in Appendix A. The various colours in Fig. 6.2 distinguish the different modelling representations used. Details of the different levels of modelling detail will be explained in the following sections.

6.3.2 *Results and Discussion*

To illustrate the effectiveness of the proposed modelling approach, three different models are created. The models used are a steady-state model, a full-dynamic model and a hybrid model. Within the steady-state model, all transmission lines are represented by algebraic equations with their dynamics ignored. The full-dynamic model has a dynamic representation of every network component with ODEs. In the hybrid model, only the key network components are modelled with ODEs, and the rest of the network is represented with algebraic equations.

Firstly, the number of state variables in the three models is calculated to illustrate the difference in the expected computational burden imposed by the models. This will be further considered by comparing the simulation times taken when the different models are used. Then, system equilibrium trajectories of the different models are calculated to show that the same operating points are produced, i.e., the same trajectories are produced by the different models. Finally, a system stability assessment, i.e., loadability calculation, along the equilibrium trajectory will be completed using bifurcation theory. Unlike the equilibrium trajectories, the system loadability assessments from the three models may vary if different bifurcations (i.e., different instability phenomena) are captured by the different models.

In this section, three different models are created. The SS-Model is a steady-state model in which Lines 1–2 and 2–3 are both modelled by algebraic equations. The Hy-Model represents a hybrid model. In the Hy-Model, the dynamics of Line 1–2 are modelled with ODEs, and the rest of the network has a static representation with algebraic equations. Referring to Fig. 6.2, the part of the network represented dynamically in the Hy-Model, i.e., Line 1–2, is coloured red. It should be noted that the three-bus test system is used to illustrate the effectiveness of the hybrid modelling approach on a small (and easy-to-understand) system, rather than the effectiveness of the entire assessment framework. Therefore, the hybrid model in Case-1 is created deliberately with manual selection of the dynamic network region rather than developed by the proposed assessment framework considering the participation of key network components. The dynamic area of the network in the Hy-Model is surrounded by a dashed line in Fig. 6.2. Finally, the Dyn-Model is a fully dynamic model, in which both Lines 1–2 and 2–3 are dynamically modelled with ODEs.

Table 6.1 summarises the overall number of state variables in each model. It reveals that the SS-Model consists of 12 state variables, the smallest of the three models. When the entire network is modelled with ODEs (Dyn-Model), the state-space model has 18 state variables. This number is reduced to 16 when the hybrid model (Hy-Model) is used. The reduced number of state variables in the Hy-Model indicates a lower-dimensional system Jacobian matrix and hence improved computational efficiency compared with the Dyn-Model.

The simulation time taken by the three different models to perform a loadability assessment is summarised in Table 6.2. Table 6.2 reveals that the steady-state model

Table 6.1 State variables for the different models in Case-1. © 2024 IEEE, reprinted, with permission, from Chen et al. [4]

Model elements	SS-Model (steady-state)	Hy-Model (hybrid)	Dyn-Model (full dynamic)
Converter	10	10	10
AC grid	2	6	8
Overall	12	16	18

requires the shortest simulation time of 3 s. The Dyn-Model takes the longest computation time of 10 s. The required time is reduced to 5 s when the hybrid model is used. Therefore, the hybrid model results in a reduction in computational burden. In combination with the results in Table 6.2, this shows that a model with a larger number of state variables comes with a longer simulation time and, hence, imposes a heavier computational burden. It should be noted that in Case-1, a small-sized power system is used. Hence, the difference in the time taken by the models is not significant. In Case-2 (shown later), for a large system, the dynamic modelling region in the hybrid model is a small proportion of the entire network, and therefore, the difference between the simulation times taken by models is greater.

The system equilibrium trajectories of the different models when the impedance of Line 1–2 Z_{12} is 0.2 p.u. are shown in Fig. 6.3. The horizontal axis is the total loading of the system, which is also the bifurcation parameter in this section, and the vertical axis represents the voltage magnitude at Bus 2.

As is shown in Fig. 6.3, the system trajectories using different models are visually identical. Considering the trajectory obtained from the full-dynamic model as the

Table 6.2 Simulation time for different models in Case-1. © 2024 IEEE, reprinted, with permission, from Chen et al. [4]

Model	Simulation time (s)
SS-Model	3
Hy-Model	5
Dyn-Model	10

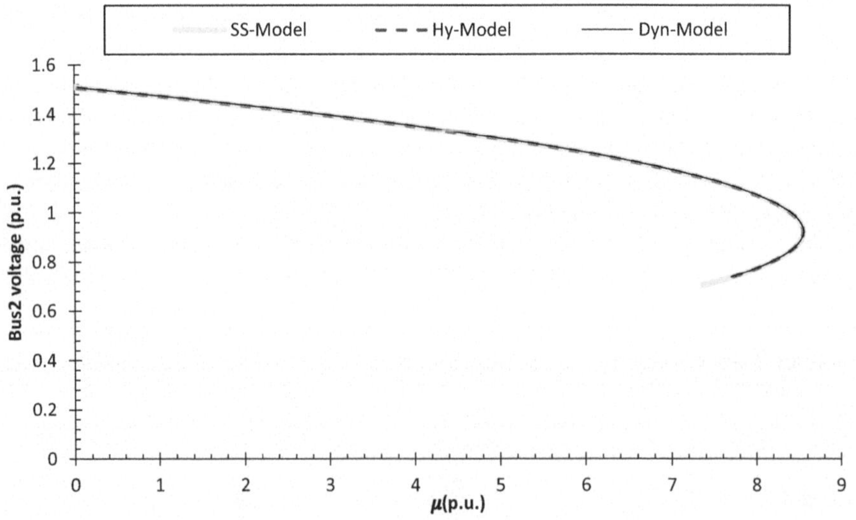

Fig. 6.3 System loadability curves obtained from different models. © 2024 IEEE, reprinted, with permission, from Chen et al. [4]

standard result, the relative root-mean-square deviations of the curve derived from the steady-state model and the hybrid model are 4.57×10^{-5} and 5.78×10^{-5}, respectively. The acceptable relative error tolerance used in this chapter is based on typical tolerances used in the continuation method defined in Chap. 2, typically around 10^{-2}. Such small errors from the steady-state model (4.57×10^{-5}) and the hybrid model (5.78×10^{-5}) could easily be due to the application of the continuation method process itself. This illustrates that the different models produce the same operating points as the loading increases.

Based on the bifurcation analysis, multiple instability mechanisms along the system trajectory will be identified as SNBs and HBs. Therefore, the system's maximum loading points identified as SNBs for the three models are essentially identical. The loadability assessment along system trajectories, including the SNB and the HB, will be investigated further in the following sections. The scenario considered will illustrate the differences in the loadability assessments when using different models.

6.3.2.1 Scenario-1

Table 6.3 summarises the loadability assessment along the system equilibrium curves for Scenario-1, where $Z_{12} = 0.2$ p.u. In Scenario-1, the system is initially stable and becomes unstable through an HB as the load increases. With the loading further increased, the system eventually encounters an SNB. From the previous discussion about system equilibrium trajectories using different models, it would be expected that the SNB values from the three models would be the same. This is confirmed in Table 6.3, where the SNB values from the three models are identical. All three models predict the SNB values at 8.56 p.u. As shown in Table 6.3, the HB locations from the three different models are also similar. The HB value using Hy-Model is 8.37 p.u., which is the same as that from the SS-Model. It can also be seen that the Dyn-Model has a similar HB value at 8.33 p.u. Considering the results from the Dyn-Model as the standard result, the relative error of the HB value from SS-Model and Hy-Model is 4.80×10^{-3}. This relative error is below the acceptable error tolerance of 10^{-2}.

Figure 6.4 details the participation factor analysis of the critical mode that causes the HBs in different models. The first ten state variables belong to the VSC, and the last six represent the network dynamics.

Table 6.3 Bifurcations along system loadability curves from different models in Case-1-Scenario-1.

Model	Locations of different bifurcations (p.u.)	
	HB	SNB
SS-Model	8.37	8.56
Hy-Model	8.37	8.56
Dyn-Model	8.33	8.56

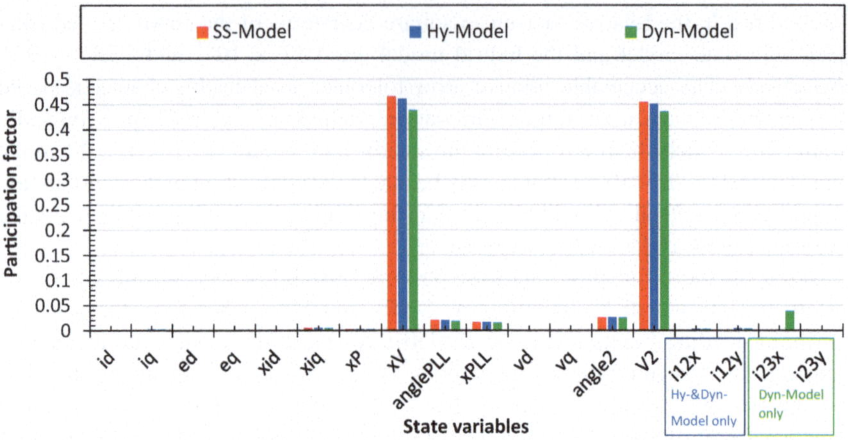

Fig. 6.4 Participations of state variables in the critical mode for Case-1-Scenario-1. © 2024 IEEE, reprinted, with permission, from Chen et al. [4]

It should be noted that in Fig. 6.4, all models show that the states associated with the voltage control of the VSC (xV) and the load voltage at Bus 2 (V_2) are the predominant participants in the critical mode. According to the Dyn-Model, the participation factor values of xV and V_2 are 0.439 and 0.437, respectively. The participation of xV and V_2 share similar values in the SS-Model (0.467 and 0.455, respectively), and in the Hy-Model (0.462 and 0.451), respectively. It can be established from these results, that all three models share a similar distribution of participation factor values. Consequently, all models can accurately determine the system stability along the loadability trajectory. Hence, if the developed methodology (in Sect. 6.2) is applied, it will identify that the dynamic area of the network is unnecessary in this scenario and the SS-Model will be used for system loadability calculations, as it not only provides an accurate result but also contains the lowest number of state variables.

The results, in this case, illustrate that though all three models accurately capture instability along the system trajectory, the full-dynamic and the hybrid models require a relatively higher-dimensional system matrix compared with the steady-state model. This scenario suggests that the inclusion of network dynamics may be unnecessary when analysing instabilities caused by interactions between SGs, motors or VSCs purely, i.e., instabilities are independent of network dynamics.

6.3.2.2 Scenario-2

Table 6.4 summarises the loadability assessment along the system trajectories when the impedance of Line 1–2 is 0.5 p.u. (2.5 times the previous case). The results from the Dyn-Model reveal that the system is initially unstable due to an HB and becomes stable as the load increases. As the load further increased, the system finally encounters an SNB. The previous discussion about system equilibrium trajectories using different models suggests that the SNB values from the three models would be the same. This is seen in Table 6.4, where the SNB values from the three models are all equal to 8.17 p.u. up to three significant figures. Table 6.4 also details the HB values obtained. It can be seen that SS-Model fails to capture the occurrence of an HB along the system trajectory. The other two models have similar results for the HB location. The Hy-Model shows that the instability will end at 8.10 p.u., and this is similar to the HB value from the Dyn-Model, which is 8.14 p.u. Considering the assessment from the Dyn-Model as the standard result, the relative error of assessment of the HB from the Hy-Model is 4.91×10^{-3}, which is smaller than the acceptable relative error range of 10^{-2}.

Figure 6.5 shows the participation factor analysis of the HBs for the different models.

The steady-state model is not included in Fig. 6.5, because it fails to capture the HB. It can be seen that the state variables of Line 1–2 (i_{12x}, i_{12y}) and state variables belonging to the VSC (e_d, e_q, x_P) participate most in the critical modes from both

Table 6.4 Bifurcations along system loadability curves from different models in Case-1. © 2024 IEEE, reprinted, with permission, from Chen et al. [4]

Model	Locations of different bifurcations (p.u.)	
	HB	SNB
SS-Model		8.17
Hy-Model	8.10	8.17
Dyn-Model	8.14	8.17

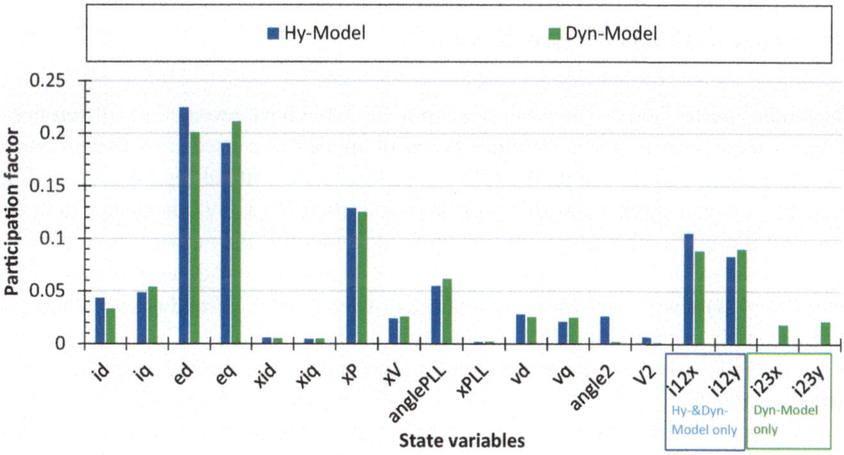

Fig. 6.5 Participations of state variables in the critical mode for Case-1-Scenario-2. © 2024 IEEE, reprinted, with permission, from Chen et al. [4]

models. The SS-Model (not shown in Fig. 6.5) does not include the dynamic representation of the critical Line 1–2. This is why it fails to capture the HBs, resulting in an inaccurate stability assessment.

According to the results obtained from the Dyn-Model, the state variables e_d, e_q participate most with participation values of 0.200 and 0.212, respectively. The state variable x_P (representing the active power control of the VSC) and the state variables of Line 1–2 (i_{12x}, i_{12y}) also participate strongly in the critical mode, with values of 0.125, 0.089 and 0.091, respectively. The Hy-Model produces similar results, and it can be established that the full-dynamic and hybrid models share a similar participation factor value distribution and, hence, both accurately assess the system stability along the loadability trajectories.

This scenario clearly illustrates that conventional modelling of the AC network with algebraic equations fails to capture certain types of instability along the system trajectory. Therefore, the steady-state model may not always be the appropriate model for loadability assessments. The full-dynamic model, on the other hand, always calculates the system loadability accurately but requires a high-dimensional system matrix, which leads to increased computational burdens. The hybrid model combines the advantages of these two models. It can assess the system loadability accurately but with a reduced computational burden. Hence, the application of the hybrid model is recommended for system loadability analysis. Additionally, it is also shown that a system may become oscillatory unstable (identified as the HB) before reaching the static maximum load point (identified as the SNB). Hence, there is a need to consider both voltage and small-disturbance stability simultaneously when assessing system loadability limits.

6.4 Case-2 IEEE 39-Bus System

The studies performed on the 3-bus system in Sect. 6.3 have revealed the differences in assessment results when different types of models are used. For that illustrative example, the proposed methodology to identify the critical regions outlined in Sect. 6.2 was not applied. In this section, a modified IEEE 39-bus system will be used and the full methodology will be applied to identify the network region to be modelled using ODEs in the hybrid model. The scenarios considered represent HBs caused by (1) network dynamics, (2) interaction between VSC and network and (3) interaction between multiple VSCs. A similar analysis process to Case-1 (the three-bus system discussed in the previous section) is used to compare the results from the different models.

The system loadability results are compared using the three different models created (i.e., the SS-Model, the Hy-Model and the Dyn-Model). With the developed assessment framework, the interactions between the VSCs and the network can be captured as bifurcations along the system loadability trajectory with a reduced computation burden.

6.4.1 *One VSC*

One of the SGs in the IEEE 39-bus system will be replaced by a VSC which shares the same capacity. The impact of dynamically modelling different regions of the network on the system stability assessment and the required number of state variables is established. This study scenario will illustrate the effectiveness of the developed methodology in a large power system.

6.4.1.1 Modification to the Test Network

A modified IEEE 39-bus system is used in Scenario-1 and is shown in Fig. 6.6. Figure 6.6 shows that the original SG at Bus 32 is replaced by a VSC which operates with $P - V_{\mathrm{ac}}$ mode. Details of the VSC parameters are given in the Appendix A.4. The dashed green line, which encapsulates the red network elements in Fig. 6.6, distinguishes the dynamic region from the static region in the hybrid model. The details of the hybrid model will be explained in the following section.

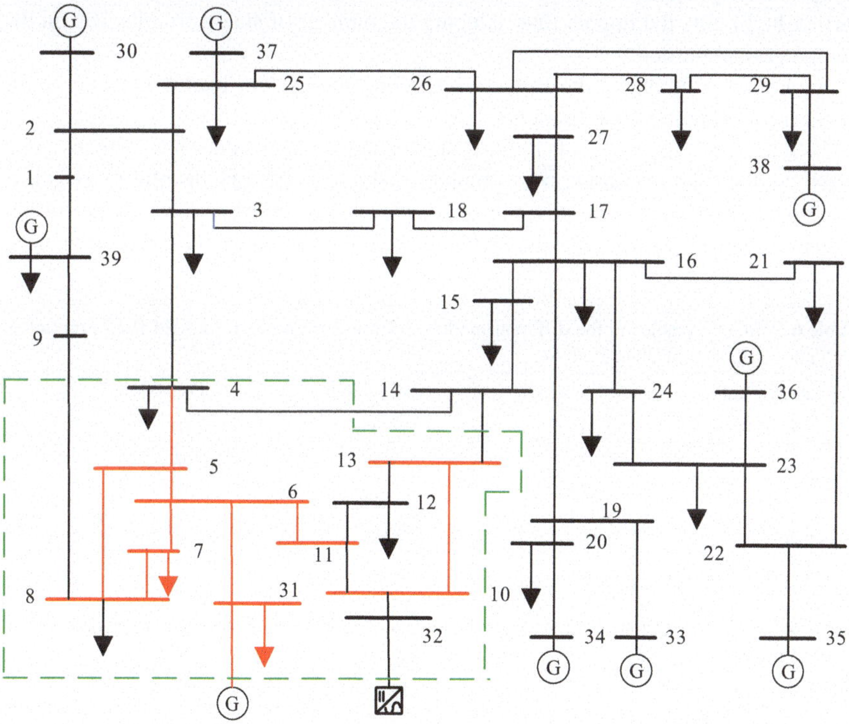

Fig. 6.6 The modified IEEE 39-bus system for Case-2-Scenario-1 (The key elements are coloured red, and the dynamic area is surrounded by a dashed line). © 2024 IEEE, reprinted, with permission, from Chen et al. [4]

6.4.1.2 Results and Discussion

For brevity, the participation of each state variable in the critical mode is not shown. According to the participation factor analysis, the critical mode is dominated by the state variables belonging to G31; the VSC; Buses 5, 6, 7, 8, 10, 11, and 13; the load at Bus 7, 31; and Lines 4–5, 5–8, 5–6, 6–7, 7–8, 6–31, 6–11, 10–11, and 10–13. Those AC components are considered as key components and are coloured red. Following the proposed assessment framework, the dynamic area of the network includes key network components and their adjacent components. Referring to Fig. 6.6, the dynamic area of the network is surrounded by a green-dashed line. According to the developed framework, the network elements within the dynamic area are modelled with ODEs, and the rest of the network is represented by algebraic equations.

Table 6.5 summarises the number of state variables in the different models. The full-dynamic model (Dyn-Model) consists of 277 state variables, which is roughly twice the number in the hybrid model.

The simulation time taken to perform a loadability assessment with the different models is shown in Table 6.6. The Dyn-Model consists of the largest number of state variables and, hence, the longest computation time (684 s). The simulation time is significantly reduced to 246 s when the developed hybrid model is used. This further highlights the impact that reducing the number of state variables has on the computational burden.

Table 6.7 summarises the bifurcations along the system loadability curves for Scenario-1 using the three models.

In Scenario-1, the system is initially stable, and all eigenvalues have negative real parts. As the load increases, the system will show oscillatory instability, which is represented as HBs. Eventually, the system reaches an SNB as the loading further increases. Table 6.7 shows that the SNB values are the same from the different models,

Table 6.5 State variables in the different models in Case-2-Scenario-1.

Model elements	SS-Model	Hy-Model	Dyn-Model
Converter	10	10	10
AC grid	0	64	206
Generators	61	61	61
Overall	71	135	277

Table 6.6 Simulation time in different models in Case-2.

Model	Simulation time (s)
SS-Model	130
Hy-Model	246
Dyn-Model	684

Table 6.7 Bifurcations along the system loadability curves from different models in Case-2-Scenario-1.

Model	Locations of different bifurcations (p.u.)	
	HB	SNB
SS-Model	2.11	2.14
Hy-Model	1.67	2.14
Dyn-Model	1.68	2.14

at 2.14 p.u. However, the HB value from the SS-Model is significantly different from the results using the other two models. As is illustrated in Table 6.7, the Hy-Model and the Dyn-Model have similar HB values of 1.67 p.u. and 1.68 p.u., respectively. It can also be seen that the SS-Model has a different HB value, at 2.11 p.u. Taking the results from the Dyn-Model as the standard assessment, the relative error of the HB value from the Hy-Model is 5.95×10^{-3}, which is smaller than the acceptable relative error range of 10^{-2}. However, the relative error for the SS-Model is 0.26, and this would result in a very inaccurate stability assessment. This error is due to the static (algebraic) representation of the key network components (i.e., the network around the VSC and G31) in the SS-Model.

From these results, it can be concluded that the proposed hybrid model can precisely assess the system loadability with a reduced number of state variables. This scenario again shows the necessity of considering multiple types of stability simultaneously when assessing the system loadability, as a system may become unstable via a small disturbance before experiencing voltage instability. Apart from that, the results also clearly demonstrate that the network's poor oscillation (non-converter-involved dynamics) may also lead to system instability, and the developed hybrid model is able to capture these important dynamics.

6.4.1.3 Impact of Considering Different Dynamic Areas of the Network

To further investigate the impact of dynamically modelling different proportions of the network on the system loadability assessment, two new hybrid models (i.e., the Hy-Model-2 and the Hy-Model-3) with different dynamic areas are built and are shown in Fig. 6.7. The Hy-Model-1 in this section represents the original hybrid model built and discussed in Sect. 6.4.1.1. In the first of these new hybrid models, Hy-Model-2, the dynamic region is expanded to include the lines and buses adjacent to those modelled dynamically in the Hy-Model-1. As Fig. 6.7 shows, the dynamic area modelled with ODEs in the Hy-Model-2 is surrounded by a dotted line. The second of the new models, Hy-Model-3, includes a dynamic region that has been reduced to exclude the lines and buses at the boundary between the dynamic and static areas in Hy-Model-1. Specifically, the previous boundary buses and the transmission

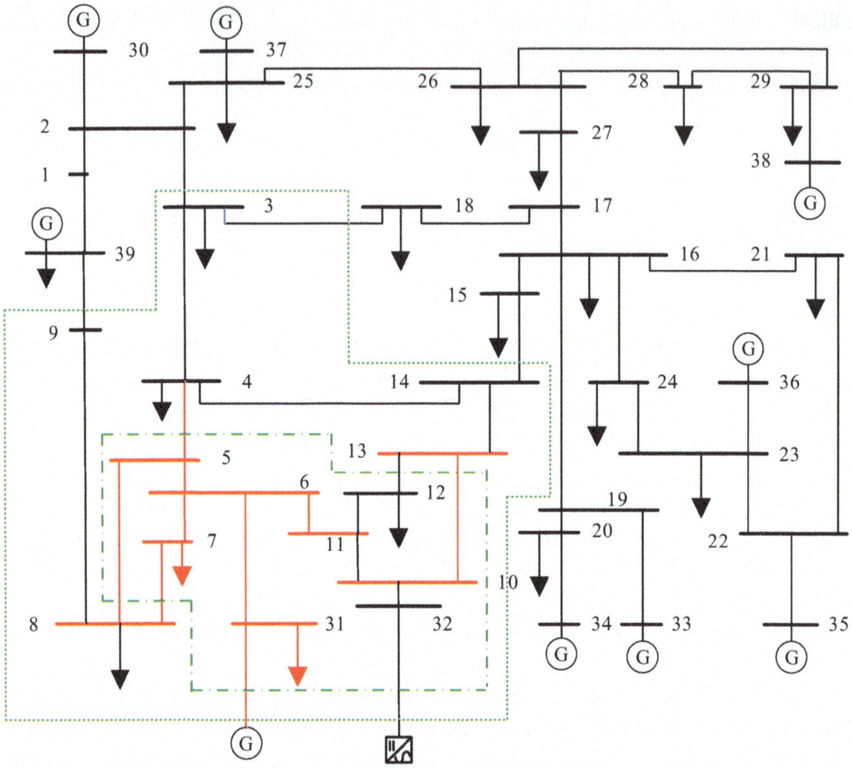

Fig. 6.7 The modified IEEE 39-bus system for different hybrid models. (The key elements are coloured red, and the dynamic area is surrounded by a dotted line for Hy-model-2 and a dashed-dotted line for Hy-model-3.)

lines connected to them will be represented by algebraic equations. The dynamic area in the Hy-Model-3 is surrounded by a dashed-dotted line, as shown in Fig. 6.7.

The numbers of state variables in the two hybrid models (shown with respect to network components) are listed in Table 6.8. Compared with 135 state variables in the Hy-Model-1 created using the developed method, the corresponding number in the Hy-Model-2 has increased to 151, due to the expansion of the dynamic area. In the Hy-Model-3, this number is reduced to 117 due to the lower number of network elements that are modelled with ODEs.

Table 6.9 summarises the loadability results obtained from Hy-Models-2 and -3. These models produce an identical SNB value of 2.14 p.u., which is the same as the previous results around SNBs shown in Table 6.7. This demonstrates that the selection of the dynamic area within the network has no impact on the shape of the system loadability curve, and that the operating points are the same in each model. As is shown in Table 6.9, the HB value of the Hy-Model-2 is the same as that using the full-dynamic model shown in Table 6.7. Hence, compared with the results using the approach proposed in the Hy-Model-1 developed from the proposed modelling

Table 6.8 State variables in different hybrid models

Model elements	Hy-Model-2	Hy-Model-3
Converter	10	10
AC grid	80	46
Generators	61	61
Overall	151	117

Table 6.9 Bifurcations along the system loadability curves from different hybrid models

Model	Locations of different bifurcations (p.u.)	
	HB	SNB
Hy-Model-2	1.68	2.14
Hy-Model-3	2.11	2.14

approach, the accuracy of the system stability analysis is improved marginally when the Hy-Model-2 is used. However, the calculation of the HB from the Hy-Model-3 is 2.11 p.u. Considering the results from the Dyn-model in Table 6.7 as standard, the relative error is 0.26, which is significant given that the acceptable error range is 10^{-2}. The significant error is primarily due to the missing key element dynamics in the Hy-Model-3.

From these results, it could be concluded that the Hy-Model-2 provides a more accurate result but at the cost of a greater number of state variables. Although in the Hy-Model-3, the number of state variables is lower, the result is unreliable and has a significant level of error. Therefore, the Hybird-Model-1 developed from the proposed methodology identifies the most appropriate proportion of the network that should be modelled dynamically for obtaining suitably accurate results, whilst taking into account the relative error level of 10^{-2}. This model provides the balance between result accuracy and computational burden.

6.4.2 Interactions Between the VSC and the Network

In the last scenario, one VSC is connected to a stiff network at which the short-circuit ratio (SCR) value is high. In this scenario (Scenario-2 of Case-2), one VSC is integrated into a weak network, where the SCR seen by the VSC is low. This scenario is included to demonstrate that the developed analysis framework accurately captures the instability due to interactions between the VSC and the network along the system loadability curve.

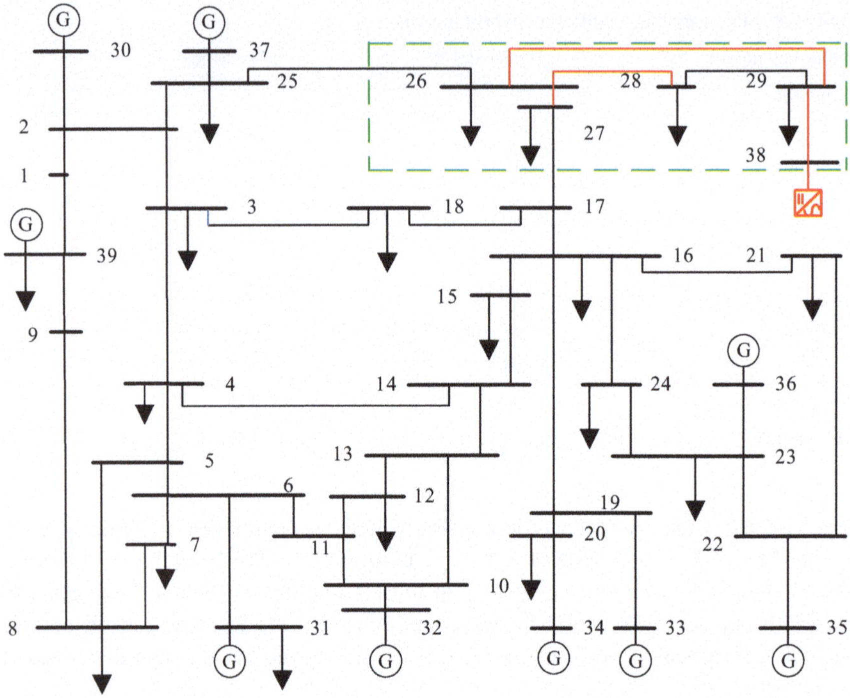

Fig. 6.8 The modified IEEE 39-bus system for Case-2-Scenario-2 (The key elements are coloured red, and the dynamic area is surrounded by a dashed line). © 2024 IEEE, reprinted, with permission, from Chen et al. [4]

6.4.2.1 Modification to the Test Network

Within this section, one VSC is connected to Bus 38 replacing the original SG. The modified IEEE 39-bus system is shown in Fig. 6.8 with VSC parameters given in Appendix A. Previous work in [14] indicates that under a weak grid condition, the interactions between the VSC and the AC network will be significant and may further degrade the system's stability. Therefore, the network impedances in the vicinity of the VSC (i.e., impedances of transmission lines 26–27, 26–28, 26–29, 28–29, and 28–38) are increased by a factor of 1.89; the loads at Buses 26, 27, 28 and 29 are increased by a factor of 1.8; and the power output of G38 is multiplied by a factor of 1.2 to exacerbate the instability due to the interaction between the VSC and the network.

6.4.2.2 Results and Discussion

According to the participation factor analysis, the critical mode represents the interaction between the VSC and the adjacent network. Correspondingly, the state variables belonging to the VSC and its adjacent network have high participation in the critical mode. The key elements identified by the proposed framework are the VSC and Lines 26–27, 26–28, 26–29, and 28–38. In Fig. 6.8, the key network components are highlighted in red. The dynamic modelling area of the network in the hybrid model is surrounded by a dashed line.

Table 6.10 details the overall number of state variables in the different models. The full-dynamic model (Dyn-Model) consists of 277 state variables, and this number decreases to 103 when the developed hybrid model is adopted, resulting in a lower computational burden.

Table 6.11 shows the bifurcations along the system loadability trajectory from the three models. The system in Scenario-2 is initially unstable and becomes stable as the loading increases. Similar to the results of SNBs in the previous scenarios, all three models produce the same SNB value, which is 1.15 p.u. The results from Dyn-Model presented in Table 6.11 reveal that the system becomes stable when the system loading increases to 0.80 p.u. The Hy-Model also has the same HB value. The SS-Model fails to capture the HB along the system loadability curve, due to the missing key network dynamics.

Scenario-2 again illustrates the effectiveness of the proposed methodology in a large system. Additionally, it can be concluded from this scenario that the proposed framework can be used to reveal the instability caused by the interactions between the VSC and the network.

Table 6.10 State variables in different models in Case-2-Scenario-2.

Model elements	SS-Model	Hy-Model	Dyn-Model
Converter	10	10	10
AC grid	0	32	206
Generators	61	61	61
Overall	71	103	277

Table 6.11 Bifurcations along the system loadability curves from different models in Case-2-Scenario-2.

Model	Locations of different bifurcations (p.u.)	
	HB	SNB
SS-Model		1.15
Hy-Model	0.80	1.15
Dyn-Model	0.80	1.15

6.4.3 Interaction Between VSCs

In Scenario-3 of Case-2, two VSCs are integrated closely into the IEEE 39-bus system to show that the developed framework has the capability of capturing instability due to the interactions between VSCs. The VSC control parameters are included in Appendix A. A similar analysis process to previous scenarios is performed in which the stability assessments and the number of state variables from various models are compared. Following this, the obtained analytical results will be verified by the time-domain simulations.

6.4.3.1 Modification to the Test Network

As is shown in Fig. 6.9, in the modified system, two VSCs are connected at Buses 30 and 37, replacing G30 and G37. To exacerbate the interactions between the VSCs (in a similar way to Scenario-2), the impedances around the VSCs are increased in this scenario. Specifically, the impedances of Lines 2–30, 2–25, and 25–37 are multiplied by a factor of 4; the power output of VSC 30 and 37 is multiplied by a factor of 2.5 and 1.2; and the power output of G38 is decreased by a factor of 0.5.

6.4.3.2 Results and Discussion

The participation factor analysis indicates that the state variables belonging to the VSCs and the network around the VSC have high participation in the critical mode. The key network components identified by the developed framework are the VSCs at Buses 30 and 37, Lines 2–30 and 25–37, Bus 2, and the load at Bus 25, which are all highlighted in red in Fig. 6.9.

The bifurcations along the system loadability trajectory from different models are shown in Table 6.12. Again, the SNB values from the three models are identical, at 1.65 p.u. The HB values calculated from the Hy-Model and the Dyn-Model are similar, at 1.23 p.u. and 1.22 p.u., respectively. However, the HB value from the SS-Model is 1.38 p.u, which is significantly different from the results from the other two models. The SS-Model fails to consider network dynamics around VSCs, which leads to a considerable error.

This scenario shows that the interactions between VSCs via the AC network can occur along the system loadability trajectory, and the developed hybrid model is capable of capturing that phenomenon.

6.4.3.3 Time-Domain Simulations

The analytical results obtained when studying this VSC–VSC interaction in the large network are further verified using time-domain simulations. Similar simulations

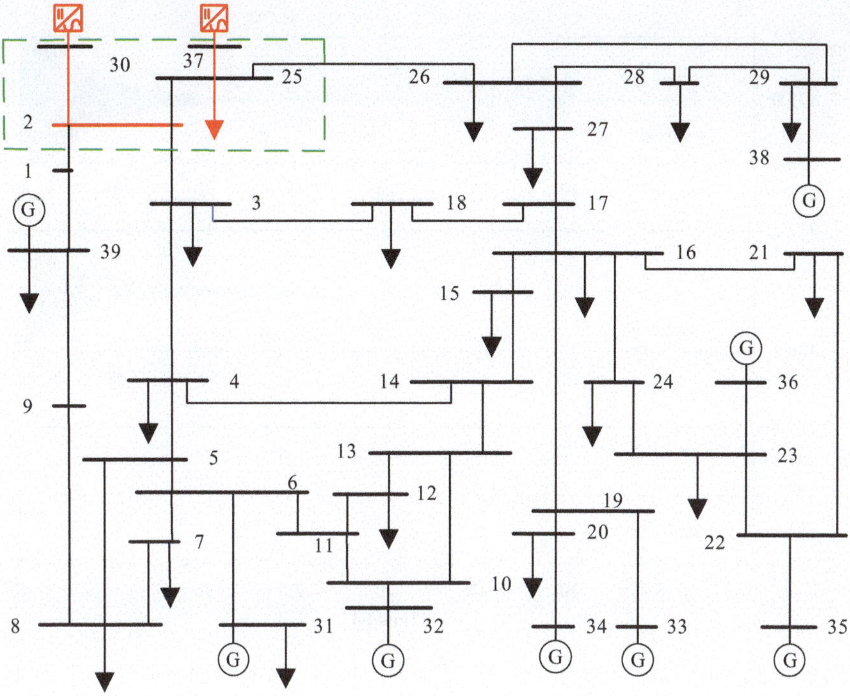

Fig. 6.9 The modified IEEE 39-bus system for Case 2-Scenario-3 (The key elements are coloured in red, and the dynamic area is surrounded by a dashed line). © 2024 IEEE, reprinted, with permission, from Chen et al. [4]

Table 6.12 Bifurcations along the system loadability curves from different models in Case-2-Scenario-3. © 2024 IEEE, reprinted, with permission, from Chen et al. [4]

Model	Locations of different bifurcations (p.u.)	
	HB	SNB
SS-Model	1.51	2.08
Hy-Model	1.23	2.08
Dyn-Model	1.22	2.08

could be produced for any of the cases and scenarios discussed in this chapter. The time-domain simulation results using the three models for a value of $\mu = 1.3$ p.u. have been produced and are shown in Fig. 6.10. Based on the analytical results shown in Table 6.12, the system is represented as unstable when the Dyn-Model and the Hy-Model are used, and stable when the SS-Model is adopted, as the HB is incorrectly represented as being located at a higher loading. A small disturbance, in which the power reference of G32 increases from 1.0 p.u. to 1.1 p.u., is injected into the system at a time of $t = 10$ s. The rotor speed of G32 (ω_{G32}) is plotted for the three different models.

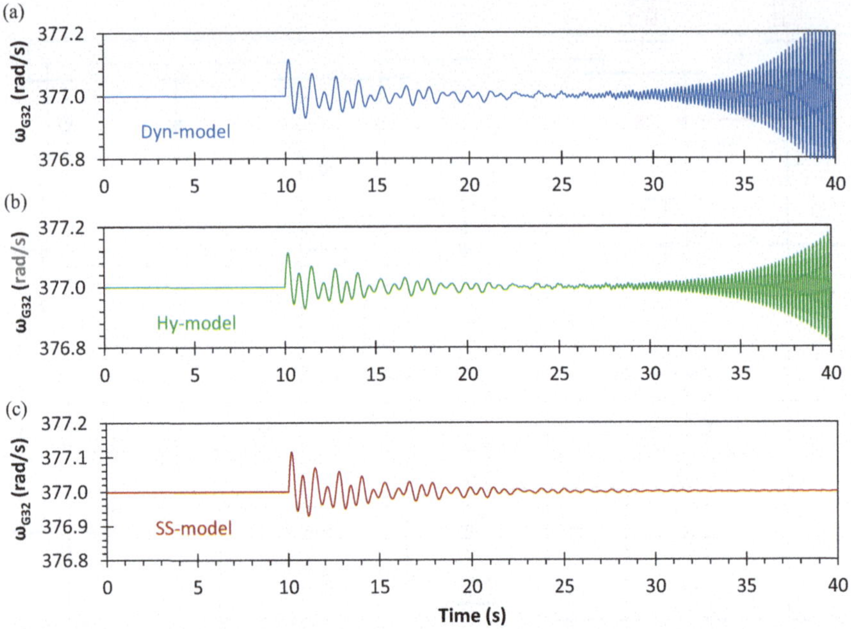

Fig. 6.10 The simulation using **a** Dyn-Model, **b** Hy-Model, and **c** SS-Model. © 2024 IEEE, reprinted, with permission, from Chen et al. [4]

The results from the Dyn-Model and the Hy-Model show that the system becomes unstable after the small disturbance, as illustrated in Fig. 6.10a and b. However, if the SS-Model is used, the rotor speed of G32 stabilises after the change in the power reference (Fig. 6.10c). It can be concluded that the time-domain simulation results verify the previous analytical results. Therefore, this demonstrates that the instability due to the interaction between VSCs could occur along the system loadability curve, and that this phenomenon could be captured by the developed hybrid model. However, the conventional steady-state model (i.e., the SS-Model) fails to capture this instability, due to the incorrect prediction of the HB location. This could lead to significant errors in loadability assessment, potentially concluding that the stability limits are higher than they are in practice.

6.5 Summary

Within this chapter, an analysis framework has been developed to efficiently analyse the system loadability, taking into consideration both small-disturbance stability and voltage stability. The proposed methodology computes the system loadability limit based on a network hybrid model combined with the application of bifurcation theory. Within the hybrid network model, the dynamic area which contains the key power components is modelled dynamically, whilst the rest of the network is represented by conventional algebraic equations. The dynamic area of the hybrid model that contains the critical network components is determined based on several techniques, including eigenvalue sensitivity and participation factor analysis. This chapter conducts comparative studies of the system loadability analysis using different network models over a wide range of scenarios. Within each scenario, three models, a steady-state model, a hybrid model and a fully dynamic model, have been developed. The initial study into the loadability assessment with different network models is presented in a modified three-bus system. The proposed analysis framework is then extensively tested on a large, IEEE 39-bus test network to show its effectiveness over a range of scenarios.

The loadability assessment conducted in the three-bus system demonstrates that a hybrid model, in which part of the network is modelled with ODEs, is able to capture system instabilities whilst reducing the computational burden. It should be noted that modelling part of the network dynamically may be unnecessary if the system instabilities are independent of network dynamics, as shown in the study performed in Scenario-1 of Case 1 in Sect. 6.3.2.1. It is necessary to establish a clear process to determine which parts of the network in the hybrid model should be modelled dynamically. The results from the modified IEEE 39-bus system reveal that if a conventional steady-state model is used, various types of instabilities, including interactions between a VSC and the network, and between VSCs via the network, may remain undetected. It has been demonstrated that the proposed analysis framework accurately captures these instabilities but with a reduced-dimensional matrix and a shorter simulation time when compared with the full-dynamic model. In general, the proposed analysis framework can identify the system loadability limits with sufficiently high accuracy but with the lowest computational burden. This framework can help system operators quickly and accurately identify power system instabilities. Furthermore, the analysis within this chapter has clearly demonstrated the necessity of considering both small-disturbance and voltage stability when calculating system loadability. This further confirms the fourth contribution of this thesis.

The methodology proposed presents the union of many of the ideas developed throughout this thesis. In Chap. 3, bifurcation theory is initially used to analyse system voltage stability. Bifurcation theory is then adopted in Chap. 4 to analyse the system feasibility region in the N-domain space, primarily considering system small-disturbance stability. Chap. 5 shows the necessity of considering small-disturbance and voltage stability when performing system loadability assessment. Based on this, the system loadability in Chap. 5 is analysed with the simultaneous consideration of multiple types of stability using bifurcation theory. Finally, this idea (the consideration of multiple types of stability when assessing loadability) is further extended in this chapter to develop an assessment framework based on a hybrid model, which can efficiently identify various system instabilities along the system loadability trajectory. The proposed framework has been tested under various network conditions and has proven to be an efficient and accurate method of capturing power instability issues caused by a range of mechanisms.

References

1. Karawita C, Annakkage UD (2010) A hybrid network model for small signal stability analysis of power systems. IEEE Trans Power Syst 25(1):443–451
2. Grdenić G, Delimar M, Beerten J (2020) Assessment of AC network modeling impact on small-signal stability of AC systems with VSC HVDC converters. Int J Electr Power Energy Syst 119:105897
3. Bidadfar A et al (2016) Power system stability analysis using feedback control system modeling including HVDC transmission links. IEEE Trans Power Syst 31(1):116–124
4. Chen Y, Preece R, Barnes M (2024) A framework for analyzing system loadability with multiple VSCs using a hybrid model. IEEE Trans Power Syst 39(1):1079–1094
5. Li P, Qi J, Wang J, Wei H, Bai X, Qiu F (2017) An SQP method combined with gradient sampling for small-signal stability constrained OPF. IEEE Trans Power Syst 32(3):2372–2381
6. Nam HK (2000) A new Eigen-sensitivity theory of augmented matrix and its applications to power system stability analysis. IEEE Trans Power Syst 15(1):363–369
7. Smed T (1993) Feasible eigenvalue sensitivity for large power systems. IEEE Trans Power Syst 8(2):555–563
8. Zhou Y, Ajjarapu V (2005) A fast algorithm for identification and tracing of voltage and oscillatory stability margin boundaries. Proc IEEE 93(5):934–946
9. Li HZ (2008) The dynamic voltage stability based on Hopf bifurcation theory in power system. In: Department electronics and eletric engineering, Shanghai Jiao Tong University, Shaihai, China
10. (Prabha) Kundur P, Balu NJ, Lauby MG (1994) Power system stability and control, 1st ed. McGraw-Hill
11. Ajjarapu V (2007) Computational techniques for voltage stability assessment and control, 1st ed. Springer
12. Kundur P, Rogers GJ, Wong DY, Ottevangers J, Wang L (1993) Dynamic Reduction. EPRI TR-102234 Project 244701, Final report. EPRI, Tech. Rep.TR-102234, Project 244701, USA, Apr-1993.
13. Wang W (2015) Operation, control and stability analysis of multi-terminal VSC-HVDC systems. Department electrical and electronic engineering, University of Manchester, Manchester, UK
14. Zhou JZ, Ding H, Fan S, Zhang Y, Gole AM (2014) Impact of short-circuit ratio and phase-locked-loop parameters on the small-signal behavior of a VSC-HVDC converter. IEEE Trans Power Deliv 29(5):2287–2296

Chapter 7
Conclusions and Future Work

7.1 Conclusions

This thesis has performed thorough stability assessments of power systems with multiple VSCs connected, using the bifurcation theory. In completing this research, bifurcation-based methodologies have been developed to assess various types of stability, including small-disturbance rotor-angle stability, resonance stability, converter-driven stability, and voltage stability.

Due to increasing loading demand, modern power systems are approaching their operational limits. Additionally, with the integration of a large number of power converters, the dynamic performance of power systems has become increasingly dependent on complex power electronic-based devices, which further affects the system's stability. The wide timescale associated with the control schemes of power electronics increases the potential for interactions with other power components. These undesirable coupling effects will lead to adverse effects on the system's stability. Additionally, potential interactions caused by the converters exist over wide-ranging timescales, which then blur the boundary between different types of stability. The combination of these factors means that it is important to assess the stability of future power systems with VSCs thoroughly considering multiple types of stability simultaneously.

To address this issue, the research presented in this thesis has adopted the bifurcation theory to first analyse each type of stability individually (i.e., voltage stability and small-disturbance stability). These studies have clearly demonstrated that bifurcation theory-based methodologies can be used to capture not only unstable oscillations caused by conventional generators but also newly emergent VSC-related instabilities. Following this, multiple types of instability are simultaneously identified and analysed along the system equilibrium trajectory using bifurcation theory methodologies during the system stability assessment. Finally, the thesis developed an analysis framework based on a hybrid network model to assess system stability efficiently. The conclusions of this research are summarised in the sections below.

Y. Chen, *Stability Assessment of Power Systems with Multiple Voltage Source Converters*, Springer Theses, https://doi.org/10.1007/978-3-031-63095-8_7

7.1.1 Voltage Stability Analysis Using Bifurcation Theory

A thorough review reveals that bifurcation theory has been extensively used to analyse the system loadability in accordance with the voltage stability of a power system. However, previous research has typically identified the system's loadability limit by studying each type of bifurcation individually. Therefore, the review clearly shows that there is a need to conduct a comprehensive evaluation of system loadability when considering multiple bifurcations and to develop an optimisation-based analysis framework which can incorporate different bifurcations simultaneously to improve the overall loadability of a system. This framework has been developed and represents the first key original contribution of this thesis. This thesis firstly updates the concept of loadability with respect to voltage stability by considering SNB, HB and LIB simultaneously. The new consideration of all three bifurcations simultaneously leads to new cases of typical loadability scenarios, as defined by the order in which bifurcations occur and their impact on the system. The new loadability scenarios defined within this work, and previously not considered, are (1) The system's maximum loadability point is an HB without reaching limits; (2) The system's maximum loadability point is an HB after encountering LIDBs.

Furthermore, the studies performed on the test networks illustrate that loadability limit boundaries are significantly different when various types of bifurcations are considered, compared to when a single type of bifurcation is examined. In general, this clearly demonstrates the necessity of simultaneously considering all relevant bifurcations when performing a loadability assessment. The proposed multi-stage, optimisation-based analysis framework contains three optimisation loops (i.e., inner, middle and outer loops), with each connected to the others in a nested structure. The proposed methodology adopts a new optimisation problem formulation which can account for different types of bifurcation simultaneously when calculating the loadability limits of a system. The studies demonstrate that the developed framework maximises the system's shortest loading margin, and subsequently optimises overall system loadability within the given parameter space. Specifically, it is shown that the proposed multi-stage framework improves the system's shortest loading margins by 9.9%, 8.4%, and 5.4% for three-bus, nine-bus and IEEE 39-bus systems, respectively.

7.1.2 Small-Disturbance Stability Analysis Using Bifurcation Theory

Beyond its application in analysing voltage stability, the review revealed that previous research has used bifurcation theory to analyse small-disturbance instability, including small-disturbance rotor-angle stability, resonance stability and converter-driven stability. Of these, converter-driven stability has received increasingly critical attention in recent years. Although stability issues related to converters have been comprehensively investigated in previous research, most efforts have been invested in investigating the instability mechanism. Due to the limitations of the approaches used, much of the existing academic research into VSC-related stability issues is limited to a small-dimensional space in which one or two parameters are varied for each scenario considered.

The second original contribution established in this thesis is the development of a methodology to efficiently identify the system feasibility region boundary in a multiple-dimensional domain. The developed methodology involves quasi-Monte Carlo simulations of operating points on the system feasibility boundary, with the initial grid conditions generated by Halton sequences. The interpolation technique is then used to construct a data-driven model based on the obtained stability points to approximate the multi-dimensional feasibility boundary of a power system. By following this proposed procedure, the developed methodology can numerically identify the system feasibility hypersurface. Illustrative studies using a small-size system demonstrate both the applicability and effectiveness of the proposed numerical approach for identifying the multi-dimensional feasibility boundary of power systems. Therefore, it has been established that the interpolated system feasibility boundary can produce results which are suitably accurate compared to those obtained using the continuation method.

Several pre-existing sensitivity studies of system feasibility boundary, based on derived mathematical equations, have typically considered only one type of bifurcation. Such techniques are no longer applicable when the system feasibility boundary is comprised of different bifurcations. Moreover, sensitivity analyses in previous research have typically been performed within a small value range of parameters and considered variations in one parameter at a time. Such results are local, as they neglect high-order cross-effects among parameters. Therefore, this thesis uses a global sensitivity technique, a partial Spearman correlation coefficient, to analyse the system feasibility boundary. The Spearman correlation coefficient is used to quantitatively assess the relationships among the parameters which form the system stability boundary to numerically identify influential parameters. Parameters with high absolute correlation values are considered to be critical factors affecting the system feasibility boundary. It is shown that compared with the outer voltage control integral gain ($k_{iV\,\mathrm{ac}}$), the impedance adjacent to the VSC (Z_{13}), PLL bandwidth (ω_{pll}), and the VSC voltage control proportional gain ($k_{pV\,\mathrm{ac}}$) are identified as influential parameters that have a significant impact on the system's feasible boundary. Additionally, the sign of the correlation index suggests the relationship between parameters. Moreover, a sensitivity study presented in a multi-VSC test system clearly reveals interactions between the VSCs and the network based on the calculated correlation index. This study represents the third contribution of this thesis.

7.1.3 Multiple Stability Analysis Using Bifurcation Theory

The existing research typically analyses each classification of stability individually. This is also true in previous multi-stability studies, in which system operating limits considering each type of stability are identified and discussed separately using individual methodology. This thesis adopts bifurcation analysis to enable a joint consideration of small-disturbance and voltage stability when assessing system loadability limits. Based on the concurrent consideration of multiple types of stability, a more comprehensive analysis of system stability is conducted. This novel research finding represents the fourth original contribution of this thesis. The studies performed on the modified IEEE 39-bus clearly reveal that a system may encounter small-disturbance instability before meeting with the voltage operational limit as the system loading increases. This confirms the necessity to consider both small-disturbance and voltage stability when assessing the system loadability.

Existing literature has extensively investigated the characteristics of different VSC outer controllers and their impact on the dynamic behaviour of systems. Most of the available research into VSC controllers primarily compares the dynamic responses of various VSC controllers within a small operating range, though a limited number of scenarios are occasionally considered. The review included highlights the need for a comprehensive assessment of VSC controllers. A thorough evaluation of the performance of different controllers across a wide range of system loadability scenarios has been completed, which represents the fifth contribution of the thesis. The assessment has shown that the PV_{ac} controller provides a larger feasible operating range when one VSC is connected to the power system, and the PV_{ac}–PV_{ac} controllers can offer the largest feasible loading margin when multiple VSCs are integrated into the system.

7.1.4 Stability Assessment Using a Hybrid Model

Due to the increasing number of power electronics-based devices, previous research has begun to incorporate the dynamics of the network by modelling network components with ODEs to capture higher frequency oscillations caused by the converter when assessing the small-disturbance stability of large power systems. It should be noted that the dynamic modelling of an entire network leads to the inclusion of a large number of state variables and, therefore, a high-dimensional system Jacobian matrix. Although this ensures accurate results, it is neither necessary nor computationally efficient to model every network component using ODEs, particularly for stability analysis of large systems. This thesis has developed an analysis framework to more efficiently assess system stability along the loadability trajectory using a hybrid network model in which the critical network area is modelled dynamically and the rest of the network is represented with algebraic equations. This framework is the sixth original contribution of the thesis.

Hybrid models developed in past research lacked a clearly defined methodology to determine which portions of the network that should be modelled with ODEs in the hybrid model. Within the hybrid modelling approach proposed within this thesis, eigenvalue sensitivity is combined with participation factor analysis to identify the critical power components and subsequently to determine the dynamic area within the hybrid model that should be modelled with ODEs. Different network models (i.e., the steady-state model, the hybrid model and the full-dynamic model) are used to evaluate the loadability limits of the system by considering instability caused by different mechanisms, including unstable dynamics of the network, interactions between the VSC and the network, and interactions between different VSCs. The comparative studies conducted using the large test network have clearly demonstrated that the proposed hybrid modelling approach is able to identify those instabilities along the system loadability trajectory accurately while reducing the computational burden. Based on the studies conducted on the modified IEEE 39-bus, the hybrid model provides suitably accurate results within an acceptable relative error of 10^{-2}, but only requires approximately one-third of the simulation time needed for the full-dynamic model that uses ODEs to model the entire network. It should be noted that the computational efficiency advantage of the hybrid model can be even more apparent when the analysis is performed in a larger power system.

7.2 Future Work

The work presented in this thesis has fulfilled the research aims and objectives stated at the outset. There still exist a number of areas within this thesis where extended studies could further develop the established ideas and models. This section discusses potential directions future work might take, categorised under the following three themes.

7.2.1 Stability Analysis Using Bifurcation Theory

This thesis provides a comprehensive assessment of system stability using local bifurcation theory-based methodologies to simultaneously consider various types of stability. This includes small-disturbance rotor-angle stability, converter-driven stability, resonance stability and voltage stability. However, other types of stability, including transient stability and frequency stability, are not considered, as they are outside the scope of this thesis. Additionally, the analysis methodologies used in this thesis (i.e., local bifurcation theory) are inapplicable for investigating the instability phenomena associated with these two types of stability. The work in [1] has demonstrated that power oscillation damping (POD) controllers, which help to improve system small-disturbance rotor-angle stability, may interact with fast frequency response (FFR), which aims to improve the system frequency stability. Additionally, a report from CIGRE suggests that the timescales of phenomena associated with transient stability and frequency stability overlap due to the introduction of converter-based devices [2]. Therefore, it would be useful for future work to incorporate these two types of stability (i.e., frequency stability and transient stability), combined with the aforementioned four types of stability simultaneously when computing the stability limits of a system. This would provide a more accurate and thorough system stability assessment. In the existing research, bifurcation theory, particularly global bifurcation theory, has been adopted to analyse power system transient and frequency stability [3–6]. For example, global bifurcation theory is combined with energy functions in [3] to analyse power system transient stability. In terms of its application in frequency stability assessment, the work in [6] employs the bifurcation theory to investigate the impact of the penetration of renewable energy resources on the frequency stability margin. Hence, it is recommended to extend the currently used local bifurcation theory to global bifurcation theory in order to conduct a comprehensive multi-stability assessment, taking into account small-disturbance stability, voltage stability, frequency stability, and transient stability simultaneously.

7.2.2 VSC Modelling

This thesis adopts bifurcation theory-based approaches to identify and analyse the interactions between VSCs and the network along the system equilibrium trajectory. The studies performed on power systems with VSCs within this thesis focus on the control interaction phenomena at a system level. Within the scope of the given study, fast electromagnetic transients (e.g., the switching functions of the VSC) can be disregarded, as they exist at a much higher frequency range and hence have negligible impact on the interaction phenomena with which the thesis is concerned. Based on this, the VSC modelling in this thesis employs a simplified average-value model (Type-6 model). The results from the studies presented here show that the VSC Type-6 model provides suitably accurate results, while significantly reducing the simulation time. However, in a practical network, beyond the interactions at the low-frequency range discussed in this thesis, VSCs can interact with other fast-response power components (including the transmission network, other power electronic-interface devices and stator dynamics of SG) at a high-frequency range. An example of such fast-dynamic interactions would be that the VSC switching function may trigger the LC resonance associated with the LCL filter, which could lead to fast-interaction, converter-driven instability [7]. To capture such high-frequency instability, a detailed equivalent MMC model, for example, the one developed in [8] that includes the energy transfer function between sub-modules, could be implemented to provide a more accurate and realistic simulation of the fast dynamic transients of VSCs. This model combined with bifurcation theory, can provide a more systematic and thorough investigation of power system stability.

Throughout this thesis, interactions between VSCs coupled with AC networks have been identified, and the impacts of these interactions on AC-side dynamics are discussed. Therefore, only the AC-side dynamics are modelled in the thesis, assuming that the VSC is connected to a stiff DC network in which DC-side dynamics are perfectly controlled by the other end of the VSC. This assumption is valid when analysing the interactions between the VSC and the network considering small-disturbance stability [9]. The comparative study included in Appendix C demonstrates the validity of this assumption. In real power systems, the AC and DC sides of the network are coupled by the VSC, and the DC-side dynamics can propagate through to the AC network, particularly when considering the system's transient stability. Although the representation of a DC network with a constant DC voltage source does not invalidate the results obtained from the small-signal interaction studies within this thesis, a more thorough representation of DC networks, as well as a detailed VSC-HVDC full-link model, could be included to accurately capture the system's DC-side dynamics. It should be noted that a lumped parameter model (i.e., nominal π model) has been widely used for DC network modelling in previous research [10]. However, this model is characterised by a typically small time constant which limits its applicability, particularly within slow transient studies of DC-side dynamics. The frequency-dependent DC cable model proposed in [11] would offer a more accurate fast dynamic response of the DC network by capturing the travelling wave effect. With a more detailed and realistic representation of the DC network, a comprehensive investigation of interactions between the VSC-HVDC and the network, and the impact of such interactions on the dynamic behaviour of the system can be completed.

7.2.3 Network Simplification Techniques

This thesis has developed an analysis framework based on a hybrid model to accurately and efficiently assess system stability along the system equilibrium trajectory. Beyond the hybrid model demonstrated here, there are other network simplification techniques that can reduce the computational burden. These include the coherency-based dynamic equivalent model [12], external Ward-type equivalency [13], and model order reduction techniques [14]. One recommendation for future studies would be to thoroughly examine the performance of the simplified networks using these network reduction techniques. Such a comparative study would evaluate the simplified network models developed from these techniques using the same test network, considering different power system issues, across a wide range of grid operating conditions, accounting for various types of stability to establish their relative performance, applicability, simplicity, and robustness in power system analysis. Such a comprehensive assessment would emphasise the benefits of using a hybrid modelling approach, and researchers would be more willing to perform system analysis based on the proposed network hybrid model.

The final direction which future research might take as a result of the findings of this thesis is to improve the developed hybrid modelling-based analysis framework. It should be noted that in the proposed methodology, the hybrid model is built based on the eigenvalue analysis of the full-dynamic model to determine the network portion that should be modelled dynamically. Therefore, a detailed model needs to be built before the construction of the hybrid model. The methodology still displays advantages in reducing simulation time in the studies presented in this thesis, because the hybrid model developed is used to analyse the entire system loadability trajectory which consists of 10^3 operating points. However, the developed methodology loses this advantage if it is adopted to analyse the small-signal stability of PE-intensive large systems at one specific operating point. Therefore, the proposed methodology would require further development so that it can build the hybrid model directly based on the network data (e.g., the network topology and system parameter settings). This may require the analysis of a large number of time-domain results conducted across numerous test networks and over a wide range of grid scenarios to create a generalised definition of the network area which requires dynamic modelling.

References

1. Asvapoositkul S, Fradley J, Preece R (2020) Incorporation of active power ancillary services into VSC-HVDC connected energy sources. Electr Power Syst Res 189:106769
2. Wilson DH et al (2016) Advances in wide area monitoring and control to address emerging requirements related to inertia, stability and power transfer in the GB power system. In: CIGRE 2016
3. Sauer PW, Pai MA, Chow JH (2017) Power system dynamics and stability: with synchrophasor measurement and power system toolbox, 2nd edn. Wiley-IEEE Press

Huang M, Peng Y, Tse CK, Liu Y, Sun J, Zha X (2017) Bifurcation and large-signal stability analysis of three-phase voltage source converter under grid voltage dips. IEEE Trans Power Electron 32(11):8868–8879

Zhang L, Sun Y, Wang Z, Tang C (2013) Analysis of bifurcation phenomena of voltage-fed inductively coupled power transfer system varying with coupling coefficient. Inf Technol J 12(6):1176–1183

6. Vinaya Mohanan VA, Mareels IMY, Evans RJ, Morton AB, Kolluri RR (2020) Stability and bifurcations in low inertia PV rich power networks. IET Gener Transm Distrib 14(25):6122–6132
7. Wang X, Blaabjerg F, Wu W (2014) Modeling and analysis of harmonic stability in an AC power-electronics- based power system. IEEE Trans Power Electron 29(12):6421–6432
8. Bergna-Diaz G, Suul JA, D'Arco S (2018) Energy-based state-space representation of modular multilevel converters with a constant equilibrium point in steady-state operation. IEEE Trans Power Electron 33(6):4832–4851
9. Tavakoli SD, Prieto-Araujo E, Sánchez-Sánchez E, Gomis-Bellmunt O (2020) Interaction assessment and stability analysis of the MMC-based VSC-HVDC Link. Energies 13(8):2075
10. DIgSILENT GmbH (2018) PowerFactory technical reference: transmission lines
11. Beddard A, Barnes M (2014) HVDC cable modelling for VSC-HVDC applications. In: 2014 IEEE PES General Meeting | Conference & Exposition, pp 1–5
12. Nath R, Lamba SS, Rao KSP (1985) Coherency based system decomposition into study and external areas using weak coupling. IEEE Trans Power Appar Syst PAS-104(6):1443–1449
13. Ward JB (1949) Equivalent circuits for power-flow studies. Trans Am Inst Electr Eng 68:373–382
14. Ghosh S, Senroy N (2013) Balanced truncation based reduced order modeling of wind farm. Int J Electr Power Energy Syst 53(1):649–655

Correction to: Stability Assessment of Power Systems with Multiple Voltage Source Converters

Correction to:
Y. Chen, *Stability Assessment of Power Systems with Multiple Voltage Source Converters*, Springer Theses,
https://doi.org/10.1007/978-3-031-63095-8

The original version of the ebook was inadvertently published with incorrect on-line descriptions of all the chapter abstracts at https://link.springer.com/ and these have been replaced. There are no changes made in any of the renditions of the publication itself. The ebook has been updated with the changes.

The updated version of this book can be found at
https://doi.org/10.1007/978-3-031-63095-8

Y. Chen, *Stability Assessment of Power Systems with Multiple Voltage Source Converters*, Springer Theses, https://doi.org/10.1007/978-3-031-63095-8_8

Appendix A
Network Data

This appendix provides the parameters required to perform dynamic studies on the test systems used within this thesis. In all cases, a system base of 100 MVA is assumed.

Three-Bus System Data

The original data is adopted from Dobson et al. [1].

Load Dynamic Parameter

The load dynamic parameters are represented in Table A.1.

Network Data

The network data for the test system is shown in Table A.2, and Bus 3 is the slack bus.

Table A.1 Load dynamic parameter in the three-bus system

K_{pw} (p.u.)	K_{pv} (p.u.)	K_{qw} (p.u.)	K_{qv} (p.u.)	K_{qv2} (p.u.)	T (s)	P_0 (p.u.)	Q_0 (p.u.)	P_1 (p.u.)
0.4	0.3	− 0.03	− 2.8	− 2.1	8.5	0.6	1.3	0

Y. Chen, *Stability Assessment of Power Systems with Multiple Voltage Source Converters*, Springer Theses, https://doi.org/10.1007/978-3-031-63095-8

Table A.2 Network parameter for the three-bus system

Y_0 (p.u.)	θ_0 (°)	E_0 (p.u.)	C (p.u.)	Y_m (p.u.)	θ_m (°)
20	– 5°	1	12	5	– 5°

Table A.3 Generator dynamic parameter data for the three-bus system

x_d (p.u.)	x_q (p.u.)	x'_d (p.u.)	x'_q (p.u.)	T'_d (s)	T'_q (s)	E_m (p.u.)	P_m (p.u.)	D (p.u.)	M (p.u.)
0.146	0.00969	0.0608	0.0608	8.96	0.31	1	1	0.05	0.3

Table A.4 AVR parameter for the three-bus system

K_a (p.u.)	T_a (s)	K_f (p.u.)	T_f (s)	K_e (p.u.)	T_e (s)
20	0.20	0.05	0.35	1	0.25

Generator Dynamic Data

The generator dynamic data is included in Table A.3.

The AVR setting for the generator is given in Table A.4.

Nine-Bus System

The full system details are adopted from Chow [2].

Line Impedance

The line impedance data for the nine-bus system is shown in Table A.5.

Load Flow Data

The load flow required to complete the load flow is included in Tables A.6 and A.7. G1 connected to Bus 1 is the slack bus.

Dynamic Generator Data

The parameters of generator dynamics are given in Table A.8.

Table A.5 Line data for the nine-bus system

From bus	To bus	***R*** (p.u.)	***X*** (p.u.)	***B*** (p.u.)	Tap ratio	Phase shift angle (°)
1	4	0	0.0576	0	0	0
4	5	0.017	0.092	0.158	0	0
5	6	0.039	0.17	0.358	0	0
3	6	0	0.0586	0	0	0
6	7	0.0119	0.1008	0.209	0	0
7	8	0.0085	0.072	0.149	0	0
8	2	0	0.0625	0	0	0
8	9	0.032	0.161	0.306	0	0
9	4	0.01	0.085	0.176	0	0

Table A.6 Load data for the nine-bus system

Bus	P_L (MW)	Q_L (MVar)	Shunt inductor G_s (MVAr)	Shunt capacitor B_s (MVAr)
1	0	0	0	0
2	0	0	0	0
3	0	0	0	0
4	0	0	0	0
5	90	30	0	0
6	0	0	0	0
7	100	35	0	0
8	0	0	0	0
9	125	50	0	0

Table A.7 Generator power output data for the nine-bus system

Bus	P_G (MW)	Q_G (MVar)	V_G (p.u.)
1	72.3	27.03	1.04
2	163	6.54	1.025
3	85	− 10.95	1.025

Table A.8 Generator dynamic parameter data for the nine-bus system

Gen	x_d (p.u.)	x_q (p.u.)	x'_d (p.u.)	x'_q (p.u.)	T'_d (s)	T'_q (s)	H (s)
1	0.146	0.0969	0.0608	0.0608	8.96	0.31	23.64
2	0.8958	0.8645	0.1198	0.1198	6	0.535	6.4
3	1.3125	1.2578	0.1813	0.1813	8.59	0.6	3.01

Table A.9 AVR parameter for nine-bus system

K_a (p.u.)	T_a (s)	K_f (p.u.)	T_f (s)	K_e (p.u.)	T_e (s)
100	0.2	0.063	0.35	1	0.314

All generators use the IEEE DCIA-type exciter with the following parameters shown in Table A.9.

IEEE 39-Bus System

Full system data is obtained from [3].

Line Impedance

The line impedance data is represented in Table A.10.

Load Flow Data

The data required to complete the load flow is given in Tables A.11 and A.12. G39 connected to Bus 39 is the slack bus.

Dynamic Generator Data

The parameters of generator dynamics are given in Table A.13.

All the generators use the DC1A-type exciter, and the associated data is given in Table A.14.

VSC System Details

The VSC-HVDC half-link model is used in this thesis. The VSC system parameters used for various case studies throughout this thesis are provided.

Table A.10 Line data for the IEEE 39-bus system

From bus	To bus	***R*** (p.u.)	***X*** (p.u.)	***B*** (p.u.)	Tap ratio	Phase shift angle (°)
1	2	0.0035	0.0411	0.6987	0	0
1	39	0.001	0.025	0.75	0	0
2	3	0.0013	0.0151	0.2572	0	0
2	25	0.007	0.0086	0.146	0	0
2	30	0	0.0181	0	1.025	0
3	4	0.0013	0.0213	0.2214	0	0
3	18	0.0011	0.0133	0.2138	0	0
4	5	0.0008	0.0128	0.1342	0	0
4	14	0.0008	0.0129	0.1382	0	0
5	6	0.0002	0.0026	0.0434	0	0
5	8	0.0008	0.0112	0.1476	0	0
6	7	0.0006	0.0092	0.113	0	0
6	11	0.0007	0.0082	0.1389	0	0
6	31	0	0.025	0	1.07	0
7	8	0.0004	0.0046	0.078	0	0
8	9	0.0023	0.0363	0.3804	0	0
9	39	0.001	0.025	1.2	0	0
10	11	0.0004	0.0043	0.0729	0	0
10	13	0.0004	0.0043	0.0729	0	0
10	32	0	0.02	0	1.07	0
12	11	0.0016	0.0435	0	1.006	0
12	13	0.0016	0.0435	0	1.006	0
13	14	0.0009	0.0101	0.1723	0	0
14	15	0.0018	0.0217	0.366	0	0
15	16	0.0009	0.0094	0.171	0	0
16	17	0.0007	0.0089	0.1342	0	0
16	19	0.0016	0.0195	0.304	0	0
16	21	0.0008	0.0135	0.2548	0	0
16	24	0.0003	0.0059	0.068	0	0
17	18	0.0007	0.0082	0.1319	0	0
17	27	0.0013	0.0173	0.3216	0	0
19	20	0.0007	0.0138	0	1.06	0
19	33	0.0007	0.0142	0	1.07	0
20	34	0.0009	0.018	0	1.009	0
21	22	0.0008	0.014	0.2565	0	0
22	23	0.0006	0.0096	0.1846	0	0
22	35	0	0.0143	0	1.025	0

(continued)

Table A.10 (continued)

From bus	To bus	***R*** (p.u.)	***X*** (p.u.)	***B*** (p.u.)	Tap ratio	Phase shift angle (°)
23	24	0.0022	0.035	0.361	0	0
23	36	0.0005	0.0272	0	1	0
25	26	0.0032	0.0323	0.531	0	0
25	37	0.0006	0.0232	0	1.025	0
26	27	0.0014	0.0147	0.2396	0	0
26	28	0.0043	0.0474	0.7802	0	0
26	29	0.0057	0.0625	1.029	0	0
28	29	0.0014	0.0151	0.249	0	0
29	38	0.0008	0.0156	0	1.025	0

VSC Dynamic Data

VSC parameter on 100 MW base with $V_{base} = 34.5$ kV is represented in Table A.15.

VSC inner current controller parameters are shown in Table A.16.

VSC outer controller parameters are shown in Table A.17.

Table A.11 Load data for the IEEE 39-bus system

Bus	P_L (MW)	Q_L (MVar)	Shunt inductor G_s (MVAr)	Shunt capacitor B_s (MVAr)
1	0	0	0	0
2	0	0	0	0
3	322	2.4	0	0
4	500	184	0	0
5	0	0	0	0
6	0	0	0	0
7	233.8	84	0	0
8	522	176.6	0	0
9	0	0	0	0
10	0	0	0	0
11	0	0	0	0
12	7.5	88	0	0
13	0	0	0	0
14	0	0	0	0
15	320	153	0	0
16	329	32.3	0	0
17	0	0	0	0
18	158	30	0	0
19	0	0	0	0
20	628	103	0	15
21	274	115	0	0
22	0	0	0	0
23	247.5	84.6	0	0
24	308.6	– 92.2	0	0
25	224	47.2	0	0
26	139	17	0	0
27	281	75.5	0	0
28	206	27.6	0	0
29	283.5	26.9	0	0
30	0	0	0	15
31	9.2	4.6	0	15
32	0	0	0	15
33	0	0	0	15
34	0	0	0	15
35	0	0	0	15
36	0	0	0	15
37	0	0	0	15

(continued)

Table A.11 (continued)

Bus	P_L (MW)	Q_L (MVar)	Shunt inductor G_s (MVAr)	Shunt capacitor B_s (MVAr)
38	0	0	0	15
39	1104	250	0	15

Table A.12 Generator power output data for the IEEE 39-bus system

Bus	P_G (MW)	Q_G (MVar)	V_G (p.u.)
30	250	150.268	1.0475
31	628.437	220.577	0.982
32	650	211.089	0.9831
33	632	113.709	0.9972
34	508	166.166	1.0123
35	650	222.895	1.0493
36	560	108.203	1.0635
37	540	9.718	1.0278
38	830	34.09	1.0265
39	1000	115.957	1.03

Table A.13 Generator dynamic parameter data for IEEE 39-bus system

Bus	x_d (p.u.)	x_q (p.u.)	x'_d (p.u.)	x'_q (p.u.)	T'_d (s)	T'_q (s)	H (s)
30	0.1	0.069	0.031	0.008	10.2	0.7	42.00
31	0.295	0.282	0.0697	0.17	6.6	1.5	30.30
32	0.2495	0.237	0.0531	0.0876	5.7	1.5	35.80
33	0.262	0.258	0.0436	0.166	5.7	1.5	28.60
34	0.67	0.62	0.0132	0.166	5.4	0.4	26.00
35	0.254	0.241	0.05	0.0814	7.3	0.4	34.80
36	0.295	0.292	0.049	0.186	5.7	1.5	26.40
37	0.29	0.28	0.057	0.0911	6.7	0.4	24.30
38	0.2106	0.205	0.057	0.0587	4.8	2	34.50
39	0.02	0.019	0.006	0.008	7	0.7	500

Table A.14 AVR parameter for IEEE 39-bus system

K_a (p.u.)	T_a (s)	K_f (p.u.)	T_f (s)	K_e (p.u.)	T_e (s)
100	0.2	0.063	0.35	1	0.314

Table A.15 VSC data

R_{vsc} (p.u.)	L_{vsc} (p.u.)	C_f (p.u.)
0.012	0.15	0.15

Table A.16 VSC inner controller parameters

$K_{p_{id}}$ (p.u.)	$K_{i_{id}}$ (p.u.)	$K_{p_{iq}}$ (p.u.)	$K_{i_{iq}}$ (p.u.)
0.5	5	0.5	5

Table A.17 VSC outer controller parameters

K_{p_P} (p.u.)	K_{i_P} (p.u.)	K_{p_Q} (p.u.)	K_{i_Q} (p.u.)	$K_{p_{V_{ac}}}$ (p.u.)	$K_{i_{V_{ac}}}$ (p.u.)
0.12	120	0.12	120	0.23	210

Appendix B
Spearman Correlation Results

This appendix details the correlation results calculated in the thesis.

Two-VSC System (Sect. 4.3.3)

The calculated partial Spearman correlation indices considering variations in the parameters of two VSCs are represented in Table B.1.

IEEE 39-Bus System (Sect. 4.4)

The correlation coefficients for a modified IEEE 39-bus system with two VSCs connected in which fourteen parameters are varied are shown in Table B.2.

IEEE 39-Bus System with One VSC (Sect. 5.2)

The correlation values between the systems parameter and feasible operating range for the IEEE 39-bus system with one VSC connected are illustrated in Table B.3.

IEEE 39-Bus System with Two VSC (Sect. 5.3)

Table B.4 displays the calculated correlation value between the different parameters and feasible ranges along the loadability curve.

Y. Chen, *Stability Assessment of Power Systems with Multiple Voltage Source Converters*, Springer Theses, https://doi.org/10.1007/978-3-031-63095-8

Table B.1 Partial Spearman correlation coefficient of each pair of bifurcation parameters for the two-VSC system in Sect. 4.3.3

	ω_{pll1}	ω_{pll2}	Z_{13}	L_{12}	k_{pP1}	k_{pP2}	$k_{pV_{ac}1}$	$k_{pV_{ac}2}$	k_{iP1}	k_{iP2}	$k_{iV_{ac}1}$	$k_{iV_{ac}2}$
ω_{pll1}	1.00	– 0.61	– 0.24	– 0.66	0.26	0.47	– 0.35	– 0.50	– 0.05	– 0.05	– 0.02	– 0.08
ω_{pll2}	– 0.61	1.00	– 0.24	– 0.75	0.29	0.54	– 0.43	– 0.58	– 0.08	– 0.02	– 0.05	– 0.08
Z_{13}	– 0.24	– 0.24	1.00	– 0.27	0.13	0.22	– 0.19	– 0.19	0.03	– 0.01	0.03	0.01
L_{12}	– 0.66	– 0.75	– 0.27	1.00	0.33	0.58	– 0.43	– 0.63	– 0.08	– 0.05	– 0.03	– 0.09
k_{pP1}	0.26	0.29	0.13	0.33	1.00	– 0.16	0.17	0.25	0.11	0.05	0.02	0.07
k_{pP2}	0.47	0.54	0.22	0.58	– 0.16	1.00	0.32	0.44	0.12	0.09	0.11	0.11
$k_{pV_{ac}1}$	– 0.35	– 0.43	– 0.19	– 0.43	0.17	0.32	1.00	– 0.26	– 0.06	0.03	0.00	0.00
$k_{pV_{ac}2}$	– 0.50	– 0.58	– 0.19	– 0.63	0.25	0.44	– 0.26	1.00	– 0.02	– 0.06	– 0.05	– 0.02
k_{iP1}	– 0.05	– 0.08	0.03	– 0.08	0.11	0.12	– 0.06	– 0.02	1.00	0.11	0.05	0.01
k_{iP2}	– 0.05	– 0.02	– 0.01	– 0.05	0.05	0.09	0.03	– 0.06	0.11	1.00	0.07	0.04
$k_{iV_{ac}1}$	– 0.02	– 0.05	0.03	– 0.03	0.02	0.11	0.00	– 0.05	0.05	0.07	1.00	0.00
$k_{iV_{ac}2}$	– 0.08	– 0.08	0.01	– 0.09	0.07	0.11	0.00	– 0.02	0.01	0.04	0.00	1.00

Table B.2 Partial Spearman correlation coefficient of each pair of bifurcation parameters for the IEEE 39-bus system in Sect. 4.4

	ω_{pll1}	ω_{pll2}	k_{pP1}	k_{pP2}	k_{pVac1}	k_{pVac2}	k_{iP1}	k_{iP2}	k_{iVac1}	k_{iVac2}	Z_{1-2}	Z_{2-3}	Z_{2-25}	Z_{25-26}
ω_{pll1}	1.00	0.07	0.08	0.09	0.08	0.04	0.10	0.07	0.06	0.09	− 0.03	− 0.03	0.12	− 0.01
ω_{pll2}	0.07	1.00	0.08	0.07	0.07	0.10	0.00	0.06	0.09	0.14	− 0.07	− 0.10	0.08	− 0.04
k_{pP1}	0.08	0.08	1.00	0.08	0.07	0.04	0.08	0.11	0.05	0.07	0.08	0.09	0.09	0.07
k_{pP2}	0.09	0.07	0.08	1.00	0.08	0.09	0.07	0.11	0.11	0.03	− 0.07	− 0.09	0.03	− 0.08
k_{pVac1}	0.08	0.07	0.07	0.08	1.00	0.09	0.10	0.08	0.09	0.08	0.05	0.06	0.09	0.05
k_{pVac2}	0.04	0.10	0.04	0.09	0.09	1.00	0.06	0.08	0.09	0.11	0.09	0.08	0.06	0.07
k_{iP1}	0.10	0.00	0.08	0.07	0.10	0.06	1.00	0.08	0.09	0.09	0.06	0.06	0.05	0.07
k_{iP2}	0.07	0.06	0.11	0.11	0.08	0.08	0.08	1.00	0.09	0.04	0.07	0.06	0.07	0.08
k_{iVac1}	0.06	0.09	0.05	0.11	0.09	0.09	0.09	0.09	1.00	0.09	0.08	0.09	0.07	0.10
k_{iVac2}	0.09	0.14	0.07	0.03	0.08	0.11	0.09	0.04	0.09	1.00	0.06	0.04	0.01	0.05
Z_{1-2}	− 0.03	− 0.07	0.08	− 0.07	0.05	0.09	0.06	0.07	0.08	0.06	1.00	− 0.79	0.12	− 0.62
Z_{2-3}	− 0.03	− 0.10	0.09	− 0.09	0.06	0.08	0.06	0.06	0.09	0.04	− 0.79	1.00	0.12	− 0.68
Z_{2-25}	0.12	0.08	0.09	0.03	0.09	0.06	0.05	0.07	0.07	0.01	0.12	0.12	1.00	0.08
Z_{25-26}	− 0.01	− 0.04	0.07	− 0.08	0.05	0.07	0.07	0.08	0.10	0.05	− 0.62	− 0.68	0.08	1.00

Table B.3 Partial Spearman correlation coefficient of each pair of bifurcation parameters for the modified IEEE 39-bus system with one VSC in Sect. 5.2

Parameter	Base case		Weak grid case
	PV_{ac}	PQ	PV_{ac}
Z_{1-2}	− 0.09	− 0.66	− 0.60
Z_{1-39}	− 0.02	0.14	− 0.03
Z_{2-3}	− 0.54	0.14	0.22
Z_{2-25}	− 0.01	− 0.02	0.10
Z_{2-30}	− 0.22	0.02	− 0.02
Z_{3-4}	− 0.80	0.00	0.01
Z_{3-18}	0.00	− 0.02	− 0.01
Z_{4-5}	− 0.75	− 0.07	− 0.04
Z_{4-14}	− 0.27	0.01	0.06
Z_{5-6}	− 0.53	0.01	0.03
Z_{5-8}	− 0.44	− 0.08	− 0.06
Z_{6-7}	− 0.62	− 0.02	0.02
Z_{6-11}	− 0.59	− 0.01	− 0.05
Z_{6-31}	− 1.00	− 0.60	− 0.43
Z_{7-8}	− 0.15	0.02	0.00
Z_{8-9}	− 0.86	− 0.09	− 0.08
Z_{9-39}	− 0.70	− 0.04	− 0.03
Z_{10-11}	− 0.37	− 0.08	− 0.09
Z_{10-13}	− 0.27	− 0.03	− 0.06
Z_{10-32}	− 0.95	− 0.11	− 0.10
Z_{12-11}	0.00	0.02	− 0.01
Z_{12-13}	− 0.01	0.04	− 0.05
Z_{13-14}	− 0.50	− 0.01	− 0.01
Z_{14-15}	− 0.63	− 0.02	− 0.04
Z_{15-16}	− 0.42	− 0.04	0.00
Z_{16-17}	0.01	0.01	0.07
Z_{16-19}	− 0.26	− 0.02	− 0.03
Z_{16-21}	− 0.07	0.00	0.01
Z_{16-21}	− 0.07	0.00	0.01
Z_{16-24}	− 0.01	0.03	0.04
Z_{17-18}	− 0.02	0.04	− 0.06
Z_{17-27}	0.00	− 0.01	0.05
Z_{19-20}	0.04	0.02	− 0.06
Z_{19-33}	− 0.18	− 0.01	0.00
Z_{20-34}	− 0.14	0.06	0.00
Z_{21-22}	− 0.24	− 0.02	− 0.02
Z_{22-23}	− 0.03	− 0.02	0.01

(continued)

Table B.3 (continued)

Parameter	Base case		Weak grid case
	PV_{ac}	PQ	PV_{ac}
Z_{22-35}	– 0.17	0.02	– 0.04
Z_{23-24}	– 0.12	0.00	– 0.02
Z_{23-36}	– 0.14	0.00	0.02
Z_{25-26}	– 0.02	0.02	– 0.02
Z_{25-37}	– 0.13	– 0.03	– 0.02
Z_{26-27}	– 0.13	– 0.01	0.01
Z_{26-28}	– 0.01	– 0.04	– 0.12
Z_{26-29}	– 0.08	– 0.04	– 0.15
Z_{28-29}	0.01	0.06	– 0.11
ω_{plli}	– 0.01	– 0.62	– 0.60
k_{pPi}	0.00	0.10	– 0.03
k_{pQi}/ $k_{pV_{ac}i}$	– 0.02	0.15	0.24
k_{iPi}	0.00	– 0.01	– 0.14
k_{iQi}/ $k_{iV_{ac}i}$	– 0.01	– 0.05	0.03

Table B.4 Partial Spearman correlation coefficient of each pair of bifurcation parameters for the modified IEEE 39-bus system with two VSCs in Sect. 5.3

Parameter	Weak grid case		
	$PQ-PQ$	$PQ-PV_{ac}$	$PV_{ac}-PV_{ac}$
Z_{1-2}	− 0.45	− 0.36	− 0.40
Z_{1-39}	0.03	0.12	0.03
Z_{2-3}	− 0.31	− 0.07	− 0.17
Z_{2-25}	0.00	0.01	− 0.01
Z_{3-4}	− 0.07	− 0.07	− 0.07
Z_{3-18}	− 0.01	0.00	− 0.03
Z_{4-5}	− 0.09	− 0.18	− 0.09
Z_{4-14}	− 0.05	0.01	− 0.06
Z_{5-6}	− 0.04	− 0.10	− 0.03
Z_{5-8}	− 0.04	− 0.09	− 0.08
Z_{6-7}	− 0.10	− 0.13	− 0.11
Z_{6-11}	− 0.02	− 0.10	− 0.05
Z_{6-31}	− 0.82	− 0.89	− 0.84
Z_{7-8}	− 0.04	− 0.06	0.01
Z_{8-9}	− 0.15	− 0.17	− 0.18
Z_{9-39}	− 0.06	− 0.09	− 0.09
Z_{10-11}	0.02	− 0.04	0.01
Z_{10-13}	− 0.07	− 0.03	− 0.07
Z_{10-32}	− 0.30	− 0.35	− 0.30
Z_{12-11}	0.00	0.01	0.05
Z_{12-13}	0.02	− 0.01	− 0.06
Z_{13-14}	− 0.13	− 0.10	− 0.06
Z_{14-15}	− 0.05	− 0.07	− 0.12
Z_{15-16}	− 0.03	− 0.04	0.00
Z_{16-17}	− 0.02	− 0.09	0.00
Z_{16-19}	0.02	0.03	0.02
Z_{16-21}	− 0.06	− 0.04	− 0.06
Z_{16-24}	− 0.01	− 0.02	0.00
Z_{17-18}	0.01	− 0.02	− 0.03
Z_{17-27}	0.02	0.06	0.01
Z_{19-20}	0.01	0.02	0.03
Z_{19-33}	0.02	− 0.03	− 0.01
Z_{20-34}	0.01	0.04	− 0.03
Z_{21-22}	− 0.05	− 0.07	− 0.03
Z_{22-23}	0.01	0.06	0.00
Z_{22-35}	− 0.04	− 0.08	− 0.01
Z_{23-24}	− 0.04	− 0.02	0.00

(continued)

Table B.4 (continued)

Parameter	Weak grid case		
	$PQ-PQ$	$PQ-PV_{ac}$	$PV_{ac}-PV_{ac}$
Z_{23-36}	0.02	− 0.01	0.02
Z_{25-26}	− 0.08	− 0.02	− 0.13
Z_{26-27}	− 0.01	0.03	− 0.02
Z_{26-28}	− 0.02	− 0.09	0.07
Z_{26-29}	− 0.07	0.00	0.01
Z_{28-29}	0.01	− 0.02	0.01
Z_{29-38}	− 0.04	0.00	0.01
ω_{plli}	− 0.26	− 0.18	− 0.22
k_{pPi}	0.02	0.05	0.02
$k_{pQi}/\ k_{pV_{ac}i}$	− 0.17	− 0.03	− 0.08
k_{iPi}	0.02	0.01	0.01
$k_{iQi}/\ k_{iV_{ac}i}$	0.01	0.01	0.03

Appendix C
Justification of Use of VSC-HVDC Half-Link Model

This appendix compares the dynamic response of a power system with a VSC-HVDC link using different DC-side network representations. Two different test network models employed in this appendix are shown in Fig. C.1, which are (a) a full-link model, where a detailed mathematical model of the entire VSC-HVDC link is included, and (b) a half-link model, where the DC-side dynamics of VSC_2 are neglected and simply represented by a constant DC voltage source.

Figure C.1a shows a typical two-terminal VSC-HVDC system connecting two AC systems [4, 5]. Such a system consists of two VSCs, two AC systems, and a DC system. The external AC networks are commonly represented by a voltage source

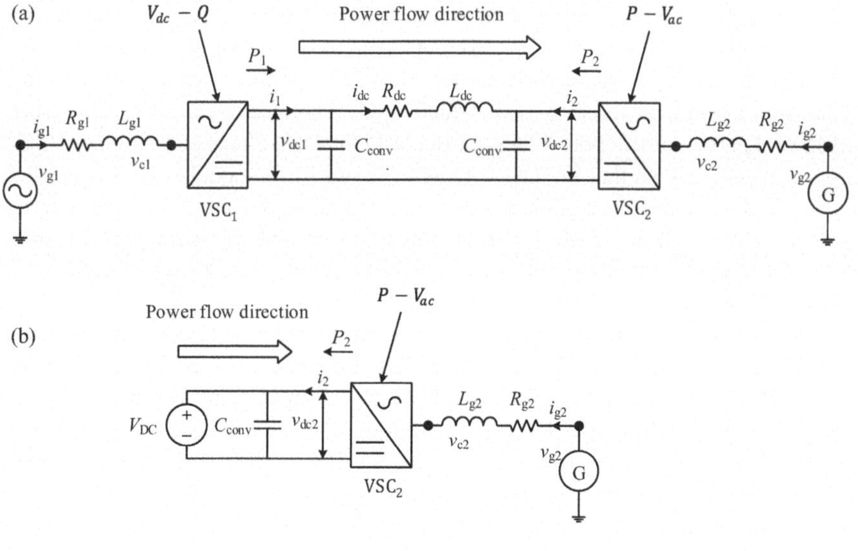

Fig. C.1 VSC-HVDC system represented with **a** full-link model and **b** half-link model

Y. Chen, *Stability Assessment of Power Systems with Multiple Voltage Source Converters*, Springer Theses, https://doi.org/10.1007/978-3-031-63095-8

Table C.1 VSC-HVDC system parameter used in Appendix C

(1) Converter basic data		
Rated power	1000 MVA	
Rated AC voltage level (phase to phase)	325.73 kV	
Rated DC voltage level	640 kV	
Equivalent converter capacitance	32 μF	
Converter resistor	0.21 Ω	
Converter reactor	6.72 mH	
(2) Nominal controller data		
Vector current controller proportional gain	K_p	0.8 p.u.
Vector current controller time constant	K_i	0.1 s
Active power controller proportional gain	K_{pP}	0.2 p.u.
Active power controller integral gain	K_{iP}	31.4 p.u.
AC voltage controller proportional gain	$K_{pV_{ac}}$	0.12 p.u.
AC voltage controller integral gain	$K_{iV_{ac}}$	50 p.u.
DC voltage controller proportional gain	$K_{pV_{dc}}$	0.2 p.u.
DC voltage controller integral gain	$K_{iV_{dc}}$	70 p.u.
Reactive power controller proportional gain	K_{pQ}	0.2 p.u.
Reactive power controller integral gain	K_{iQ}	31.4 p.u.

(for VSC_1) or a synchronous generator (for VSC_2) in series with the grid equivalent impedance. In a typical configuration of the VSC-HVDC link, the power flows from VSC_1 to VSC_2 [5]. In such a VSC-HVDC system, it can be assumed that VSC_1 operates with the $V_{dc} - Q$ outer controller and VSC_2 operates with the $P - V_{ac}$ outer controller. To simulate the scenario in which VSC_2 is connected to a stiff DC network, all control parameters of both VSCs, particularly the DC voltage outer loop of VSC_1, are tuned properly so that the DC voltage in the full-link model is well controlled by the VSC_1 station. It should be noted that the same system parameter settings shown in Table C.1 are used in both models to ensure the comparability of results. All the EMT simulations within this appendix are performed using DIgSILENT PowerFactory 2022.

The EMT simulation results obtained when using different HVDC system representations are illustrated in Fig. C.2. A small disturbance, in which the power reference of VSC_2 increases from 0.30 p.u. to 0.33 p.u., is injected into the system at time $t = 5$ s. Figure 10 shows (a) the power output of VSC_2 at the PCC and (b) phase A currents flowing out of the VSC for both models. The results using the VSC-HVDC full-link and half-link models are represented by a solid orange line and a dashed black line, respectively. It is clear from Fig. C.2a that the VSC output power from both models increases from 300 to 330 MW after the change in the reference power setting. Figure C.2a clearly illustrates that the power output of VSC_2 calculated from the half-link model accurately overlaps with that produced using the full-link model,

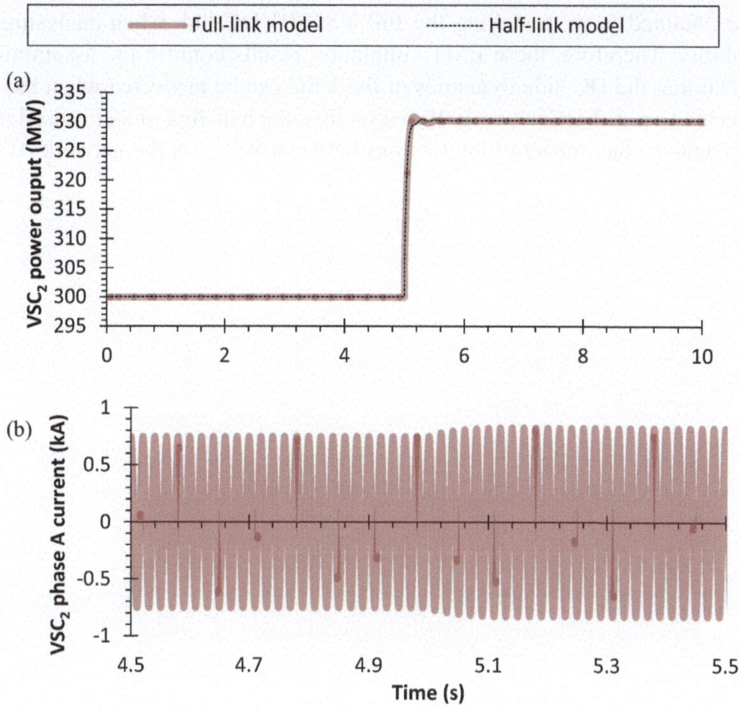

Fig. C.2 VSC-HVDC system dynamic responses using **a** VSC-HVDC full-link and **b** half-link models

except for oscillations which last a few milliseconds after the disturbance is injected. Within the full-link model, the DC voltage is controlled by VSC_1 at the other end of the HVDC link, while in the half-link model, the VSC is directly connected to a constant DC voltage source. Hence, the full-link model simulates the power system transient behaviour slightly better than the half-link model. Although the difference between initial dynamics using two different models exists, it is not critical in this study as the thesis focuses on system behaviours in a quasi-steady-state timeframe. Specifically, the system stability information at steady-state is of more interest rather than the transient behaviour after the disturbance is injected. The relative difference between the peak values of the two models with respect to the power output from the full-link model is 0.15%. This clearly demonstrates a good agreement between the results when the two models are used.

Additionally, the high degree of accuracy of results from the half-link model can be further seen from the almost perfectly overlapping phase A current curves in Fig. C.2b. Taking the curve produced from the full-link model as a baseline, the relative root-mean-square error (RMSE) of the current output using the half-link model is 2.53×10^{-8}. It is evident from Fig. C.2 that the half-link model can thus be used to provide system dynamic responses that are suitably accurate compared to those obtained from modelling the full VSC-HVDC link when analysing AC-side stability. Therefore, these EMT simulation results confirm the assumption in Sect. 2.2.6 that the DC-side dynamics of the VSC can be neglected when the VSC is connected to a stiff DC network. Based on this, the half-link model is used in this thesis to analyse the impact of interactions between VSCs on the network AC-side dynamics.

References

1. Dobson I, Chiang HD, Thorp JS, Fekih-Ahmed L (1988) Model of voltage collapse in electric power systems. In: Proceedings of the IEEE conference on decision and control, pp 2104–2109
2. Chow JH (1982) Time-scale modeling of dynamic networks with applications to power systems, 1st edn. Springer, Berlin, Heidelberg
3. Rogers G (2000) Power system oscillations, 1st edn. Springer, New York, NY
4. Huang Y, Yuan X, Hu J, Zhou P (2015) Modeling of VSC connected to weak grid for stability analysis of DC-link voltage control. IEEE J Emerg Sel Top Power Electron 3(4):1193–1204
5. Morris JF, Ahmed KH, Egea-Alvarez A (2021) Analysis of controller bandwidth interactions for vector-controlled VSC connected to very weak AC Grids. IEEE J Emerg Sel Top Power Electron 9(6):7343–7354

Y. Chen, *Stability Assessment of Power Systems with Multiple Voltage Source Converters*, Springer Theses, https://doi.org/10.1007/978-3-031-63095-8

The manufacturer's authorised representative in the EU is Springer Nature Customer Service Centre GmbH, Europaplatz 3, 69115 Heidelberg, Germany. If you have any concerns regarding our products, please contact ProductSafety@springernature.com

Printed and bound by CPI Group (UK) Ltd, Croydon, CR0 4YY
01/10/2026
02225966-0008